T0144154

Fields of Practice and Applied Solutions within Distributed Team Cognition

Fields of Practice and Applied Solutions within Distributed Team Cognition

Edited by Michael D. McNeese, Eduardo Salas, and Mica R. Endsley

CRC Press is an imprint of the
Taylor & Francis Group, an **informa** business

First edition published 2021
by CRC Press
6000 Broken Sound Parkway NW, Suite 300, Boca Raton, FL 33487–2742

and by CRC Press
2 Park Square, Milton Park, Abingdon, Oxon, OX14 4RN

CRC Press is an imprint of Taylor & Francis Group, LLC

Library of Congress Cataloging-in-Publication Data
Names: McNeese, Michael, 1954– editor. | Salas, Eduardo, editor. | Endsley, Mica R., editor.
Title: Fields of practice and applied solutions within distributed team cognition / edited by Michael McNeese, Eduardo Salas, and Mica R. Endsley.
Description: First edition. | Boca Raton, FL : CRC Press, 2020. | Includes bibliographical references and index.
Identifiers: LCCN 2020013347 (print) | LCCN 2020013348 (ebook) | ISBN 9781138626003 (hardback) | ISBN 9780429459542 (ebook)
Subjects: LCSH: Teams in the workplace. | Organizational behavior.
Classification: LCC HD66 .F544 2020 (print) | LCC HD66 (ebook) | DDC 302.3/5—dc23
LC record available at https://lccn.loc.gov/2020013347
LC ebook record available at https://lccn.loc.gov/2020013348

ISBN: 978-1-138-62600-3 (hbk)
ISBN: 978-0-429-45954-2 (ebk)

Typeset in Times
by Apex CoVantage, LLC

Dedication

Just over four years ago I was sitting in an uncomfortable chair listening to the infusion pump load chemicals into Judy (my wife) to combat ovarian cancer. It was hard to concentrate on the ideas I was putting together and my mind wandered incessantly. I was on sabbatical for the fall semester from Penn State. Little did I know that I would be sitting in Mount Nittany Medical Center taking care of Judy as she went through chemotherapy. Fortunately, I was able to devote time as a caregiver simultaneous with focusing on academic goals set for myself while on sabbatical. One of those goals was to create an interdisciplinary handbook that would examine a spectrum of contemporary topics within the cross sections of distributed cognition and team cognition as they applied within given contexts of use. The result of that seed concept is this Handbook of Distributed Team Cognition. *I would like to dedicate this handbook to Judy for the strength, courage, character, and positive outlook she showed during her time of confronting cancer and fighting to get healthy. By the grace of God, she is now cancer-free and healthy again. We just had our 40th anniversary of marriage this May of 2020—we celebrate our mutual love and continued commitment to support each other in every way possible. We look at life as an adventure that may present unexpected yet beneficial experiences that make us better humans. I love you.*

Michael D. McNeese

Contents

Preface

Many different cognitive research approaches have been generated to explore what is meaningful and resonant in fields of practice where mutual teamwork is present and emergent. Results have shown subtle yet significant findings on how humans actually work together and when they transition from their own individual roles and niches into elements of teamwork and even "team-to-team" work. Likewise, studies employing fields of practice have revealed highly informative data as to how humans (groups, teams) work with unique tools and technologies and adapt them for specific situations, and concomitantly how tools and technologies can be designed to fit with individual differences and teamwork. With many fields of practice utilizing various computational functions—often situated within online systems and data—experience, practice, and learning are highly distributed across objects, tools, interfaces, environments, and people in dynamic and unique ways. This volume, as part of overall set comprising the *Handbook of Distributed Team Cognition*, represents and offers an understanding of specific fields of practice that (1) are discovered using differing methodological approaches, (2) employ innovative aspects of computing and technology within a situated environment, and (3) are highly coupled with data/information science products about how cognition emerges and is applied and practiced in the real world. Study of real-world environments provides continuously refined specificities of "knowing" wherein experts can be delineated from novices.

Technology is absolutely changing the world we live in and how we come to define what we do and how we do it. In some cases, it produces negative impacts, but in other cases it amplifies the bandwidth of what is possible and increases value and worth of our well-being. This is certainly the case when one examines the role cognition has in teamwork as it is distributed across dynamic contexts.

The provision of computational support and technological innovation absolutely changes specific areas of cognition such as pattern recognition, attention, memory, learning, situation awareness, decision making, problem solving, judgment, perception, calculation, creativity, language, and imagination as well as how these areas are distributed across teams, ecologies, and social systems. One can see this just in the area of medical practice where new support technologies and medical informatics improve the diagnostic capabilities of the medical team and enhance the quality and length of life for people undergoing many types of illness and distress. However, as we have seen in the past, if technology is designed without human considerations in mind—or if it is used in the wrong way without proper training—then errors, failures, or even loss of life may occur. The contemporary study and comprehension of distributed team cognition cannot be constrained and reconciled in an isolated vacuum. Rather it must be considered in the broader context within which it occurs—which in part is governed by the powers of technologies, information science, and computation. Cognition not only takes place in our minds, but it is distributed across time, history, context, culture, and place, and with the realization that technologies creates new experiences of *use*.

Nothing is more obvious than the example of how the *use* of smartphones changes the way we assimilate, process, and create information through new forms of interaction with many people distributed across various situations. Many types of collaborative interactions are now highly interdependent with technological applications that exist in our phones or with other kinds of computational devices that in turn impact how teams and/or groups of people seek, exchange, share, adapt, store, process, publish, and communicate information repeatedly. Social media *use* intricately crisscrosses cognition and is distributed across many levels and layers of networks for consumption.

Distributed team cognition today is interlaced with forms of technological substrates that influence and connote our interpretation of what cognition means within situated contexts for given fields of practice. Human-technological affordances produce interactions that now can be distributed across many realms simultaneously (e.g., remote synchronous work) or that allow work to be interrupted but continued as appropriate when needed (e.g., online asynchronous command and control processes). These new levels of affordances are changing societal notions of organization, intelligence, access, privilege, privacy, security, warfare, emergency response, and entertainment, to name a few. Advancing computation to viable forms of discovery (e.g., machine learning) is enabling significant progress in systems design, collective consciousness, and social awareness. For example, recent work in human-autonomous systems is redefining everyday experience. Take the example of self-driving cars and what this means for such elements of cognition as attention, safety, relaxation, and quality of life. Distributed team cognition now encompasses newer growth expansion in such specialties as the internet of things, data science, crowd sourcing, citizen science, intelligent agents and robotics, facial recognition, enterprise architectures, social networks, augmented reality, visualization, mobile wearable computing, and information sense surrounds, to name a few, which are predicated through the advancement of technologies, designed for humans, and enabled for fields of practice.

Michael D. McNeese
February 2020

SPECIAL NOTIFICATION

This book was produced during the outbreak of the COVID-19 crisis that caused many people around the world by necessity to participate in distributed team cognition in their everyday lives. Many of us were required to be separated by distance to avoid the spread of the virus, hence collaborative work/meetings, joint entertainment, church services, and other activities were conducted through the use of distributed tools, technologies, and apps. As such, the topics within this book are highly relevant for the times we live in and are experiencing. Technologies such as Zoom and Skype facilitated connectedness, teamwork, and social awareness that enabled life to continue in the best possible way. As we adapt to the circumstances of this virus, perhaps many elements of distribute team cognition will be inculcated as part of our permanent culture/society. As we face the summer of 2020, the trajectory of COVID-19 is uncertain and indeterminate. We wish all those affected and impacted by COVID-19 the best path forward.

Editors

Michael D. McNeese is a Professor (Emeritus) and was the Director of the MINDS Group (Multidisciplinary Initiatives in Naturalistic Decision Systems) at the College of Information Sciences and Technology (IST), The Pennsylvania State University, University Park, PA. Dr. McNeese has also been a Professor of Psychology (affiliated) in the Department of Psychology, and a Professor of Education (affiliated) in the Department of Learning Systems and Performance, at Penn State. Previously, he was the Senior Associate Dean for Research, Graduate Studies, and Academic Affairs at the College of IST. Dr. McNeese also served as Department Head and Associate Dean of Research and Graduate Programs in the College, and was part of the original ten founding professors in the College of IST. He has been the principal investigator and managed numerous research projects involving cognitive systems engineering, human factors, human-autonomous interaction, social-cognitive informatics, cognitive psychology, team cognition, user experience, situation awareness, and interactive modeling and simulations for more than 35 years. His research has been funded by diverse sources (NSF, ONR, ARL, ARO, AFRL, NGIA, Lockheed Martin) through a wide variety of program offices and initiatives. Prior to moving to Penn State in 2000, he was a Senior Scientist and Director of Collaborative Design Technology at the USAF Research Laboratory (Wright-Patterson Air Force Base, Ohio). He was one of the principal scientists in the USAF responsible for cognitive systems engineering and team cognition as related to command and control and emergency operations. Dr. McNeese received his Ph.D. in Cognitive Science from Vanderbilt University and an M.A. in Experimental-Cognitive Psychology from the University of Dayton, was a visiting professor at The Ohio State University, Department of Integrated Systems Engineering, and was a Research Associate at the Vanderbilt University Center for Learning Technology. He has over 250 publications in research/application domains including emergency crisis management; fighter pilot performance; pilot-vehicle interaction; battle management command, control, communication operations; cyber and information security; intelligence and image analyst work; geographical intelligence gathering, information fusion, police cognition, natural gas exploitation, emergency medicine; and aviation. His most recent work focuses on the cognitive science perspectives within cyber-security utilizing the interdisciplinary Living Laboratory Framework as articulated in this book.

Eduardo Salas is the Allyn R. and Gladys M. Cline Chair Professor and Chair of the Department of Psychological Sciences at Rice University. His expertise includes assisting organizations, including oil and gas, aviation, law enforcement and healthcare industries, in how to foster teamwork, design and implement team training strategies, create a safety culture and minimize errors, facilitate learning and training effectiveness, optimize simulation-based training, manage decision making under stress, and develop performance measurement tools.

Dr. Salas has co-authored over 480 journal articles and book chapters and has co-edited 33 books and handbooks as well as authored one book on team training. He is a Past President of the Society for Industrial and Organizational Psychology (SIOP) and the Human Factors and Ergonomics Society (HFES), and a fellow of the American Psychological Association (APA), Association for Psychological Science, and HFES. He is also the recipient of the 2012 Society for Human Resource Management Losey Lifetime Achievement Award, the 2012 Joseph E. McGrath Award for Lifetime Achievement for his work on teams and team training, and the 2016 APA Award for Outstanding Lifetime Contributions to Psychology. He received his PhD (1984) in industrial/organizational psychology from Old Dominion University.

Mica R. Endsley is the President of SA Technologies, a cognitive engineering firm specializing in the development of operator interfaces for advanced systems, including the next generation of systems for military, aviation, air traffic control, medicine, and power grid operations. Previously she served as Chief Scientist of the U.S. Air Force in where she was the chief scientific adviser to the Chief of Staff and Secretary of the Air Force, providing assessments on a wide range of scientific and technical issues affecting the Air Force mission. She has also been a Visiting Associate Professor at MIT in the Department of Aeronautics and Astronautics and Associate Professor of Industrial Engineering at Texas Tech University. Dr. Endsley is widely published on the topic of situation awareness and decision making in individuals and teams across a wide variety of domains. She received a PhD in industrial and systems engineering from the University of Southern California. She is a past president and fellow of the Human Factors and Ergonomics Society and a fellow of the International Ergonomics Association.

Contributors

Emma Allen*
Deloitte and School of Information Studies
Syracuse University
Syracuse, New York

Matthew S. Arbogast
United States Military Academy
West Point, New York

Kevin B. Bennett
Department of Psychology
Wright State University
Dayton, Ohio

Ann M. Bisantz
Department of Industrial and Systems Engineering
University at Buffalo
Buffalo, New York

Lora A. Cavuoto
Department of Industrial and Systems Engineering
University at Buffalo
Buffalo, New York

Beomkyu Choi
Department of Literacy and Technology
Grand Valley State University
Grand Rapids, Michigan

Michael D. Coovert
Department of Psychology
University of South Florida
Tampa, Florida

Ewart J. de Visser
Warfighter Effectiveness Research Center
United States Air Force Academy
Colorado Springs, Colorado

John M. Flach
Mile Two, LLC
Dayton, Ohio

Peter K. Forster
College of Information Sciences and Technology
The Pennsylvania State University
University Park, Pennsylvania

Jonna M. Kulikowich
Department of Educational Psychology, Counseling, and Special Education
The Pennsylvania State University
University Park, Pennsylvania

Diana Kusunoki
User Experience and Research
White Ops, Inc.
New York, New York

Jean Brittain Leslie
Center for Creative Leadership
Greensborough, North Carolina

Margaret M. Luciano
Department of Management and Entrepreneurship
Arizona State University
Tempe, Arizona

Vincent Mancuso
Cyber Operations and Analysis Technology Group
MIT Lincoln Laboratories
Lexington, Massachusetts

John E. Mathieu
Department of Management
University of Connecticut
Storrs, Connecticut

* Current affiliation. Work done while at Syracuse University.

Sarah McGuire
Cyber Operations and Analysis Technology Group
MIT Lincoln Laboratories
Lexington, Massachusetts

Michael D. McNeese
College of Information Sciences and Technology
The Pennsylvania State University
University Park, Pennsylvania

Carsten Østerlund
School of Information Studies
Syracuse University
Syracuse, New York

Steve Sawyer
School of Information Studies
Syracuse University
Syracuse, New York

Sarika Sharma
School of Information Studies
Syracuse University
Syracuse, New York

Jaime Snyder
Information School
University of Washington
Seattle, Washington

Samantha Weirman
College of Information Sciences and Technology
The Pennsylvania State University
University Park, Pennsylvania

Matt Willis
Oxford Internet Institute
Oxford University
Oxford, United Kingdom

Michael F. Young
Department of Educational Psychology
University of Connecticut
Storrs, Connecticut

Zhan Zhang
School of Computer Science and Information Systems
Pace University
New York, New York

Primer (Introduction)

In volumes 1 and 2 we have seen the portrayal of distributed team cognition through many angles that reflect different spectrums and highlight creative ideas while unifying shared knowledge on various subjects. People have worked together in teams to address difficult and challenging problems for a long time. In many cases, teamwork has resulted in positive gains and effective work whereas in other cases errors have been encountered and performance has been limited. As work has become more cognitive and distributed, and as information has become ubiquitous, the role of technological innovation has changed the nature of team cognition dramatically. Even the differentiation between groups, teams, and collectives has become blurred. Within volumes 1 and 2, historical traces and pathways have been provided and documented, actual experimental studies have revealed cogent findings, modeling techniques have been used to expand detailed levels of scientific prediction, and new forms of measurement and analysis have led researchers to distill new insights, build and construct more in-depth knowledge, and suggest user interactions with technologies and their relationships with distributed team cognition. Now volume 3 will examine some of the solutions that have been suggested within distributed team cognition with specific reference to the contexts they are designed to enhance. Fields of practice are important laboratories where specific activities and relationships can foment and where distributed information has the possibility to amplify or confuse dependent on how it is used. This last volume of the handbook looks at how specific applications of distributed team cognition is manifest in the real world, providing successful benefits to people at work.

The first chapter is presented by Mathieu, Leslie, and Luciano and looks at global teamwork where teams interact and perform virtual work wherein communication and collaboration tools facilitate operations. This research focuses on the development of shared mental models in virtual work, identifying the challenges and paradoxes that need to be addressed. The chapter is axiomatic at integrating theory, methods, measures, and outcomes associated with the study of 50 virtual teams from a variety of industries and geographic areas. Results suggest that shared mental models can enhance team effectiveness. As part of their results they consider the impact of cultural differences in global teamwork.

Chapter 2 by Sawyer, Snyder, Willis, Sharma, Østerlund, and Allen provides an intriguing and interdisciplinary look at the cyberinfrastructure increasingly being used by distributed social science collaborators. The chapter looks at digital resources as they are used by practicing social scientists who are working at distance. Through interviews of these scientists the authors uncover how adaptive design practices produce digital arrangements that lead to a stabilized sociotechnical infrastructure. An underlying theme that is important is that current notions of cyberinfrastructure underplay the importance and role of social negotiation in distributed collaborative scientific work.

Chapter 3 by Young, Kulikowich, and Choi looks at the area of learning through the use of collaborative board games and provides a situated cognition description of learning from game play. The chapter is a great example of applying an ecological framework to understand how distributed cognition and learning comes about through interactive play. It provides a basis for considering individual and team play and how interactions lead to distributed cognition. Specific examples of collaborative games are provided and utilized as a basis to measure the development of dynamic distributed team cognition.

Chapter 4 by Forster and Weirman takes a unique view that technology in the form of Twitter has led to situation awareness for specific social events, in particular social unrest and the resulting Twitter activity that precedes action. Their research analyzes social tweets and how social situation awareness developed within the conflicts associated with the Unite the Right rally. The relationship between Twitter data and mobilization of groups is assessed. This research is important as it assesses the impact of a current technology (Twitter) on the emergence of social and individual situation awareness across time, showing that an online technology can influence social actions. The chapter is also important from a methodological perspective as it provides the means to understand and analyze a significantly large amount of Twitter data in order to see how distributed team cognition arises over time.

Chapter 5 by Zhang and Kusunoki is a valuable look at how teams work within in the medical domain and how team situation awareness comes about. The chapter provides a nice review of situation awareness within teams with particular emphasis on collocated, distributed, synchronous, and asynchronous work. It also delineates the use of particular technologies that provide solutions to team situation awareness problems and issues. Finally the chapter applies and utilizes activity theory as a means to understand time-critical and interdisciplinary teamwork, wherein specific technologies may address different kinds of awareness. The medical domain is one of the most critical areas where distributed team cognition comes into play and where much technological advancement can improve cooperative work.

Chapter 6 by Mancuso and McGuire addresses another critical practical area of concern for distributed team cognition, cybersecurity, which is omnipresent in many fields of practice, organizations, and businesses. This important area is one that has not been viewed from a human or team-centered perspective in most cases. Therein, this chapter is especially valuable in that it first reviews the dynamic nature of cybersecurity and the complex requirements that require teamwork perspectives. The chapter gives a thorough review of how individuals, teams, and technologies intersect to address the issues and challenges inherent within cybersecurity applications. The authors then go on to look at unique considerations of cybersecurity as a sociotechnical system and what this means for design and technology development. In particular, they look at human-machine teaming elements which will increasingly be utilized within this domain.

Chapter 7 by Cavuoto and Bisantz explores the role of distributed team cognition within the field of practice of human-robot manufacturing teams. Again, this is a dynamic and unique domain to look at cooperation among humans and robots and how coordination, cooperation, and shared information develops over time. As is true for many domains where humans team with agents and/or robots there is much opportunity for errors and even failure. How engineering and designers create systems that can work reliably to accomplish outstanding goals is a critical consideration for this handbook. The work presented draws upon theories and methodologies within human factors and specifically addresses topics such as situation awareness, trust, communication, reliable work, and function allocation. These topics are essential to deal with in many domains that incorporate distributed team cognition that requires integration of humans with technology team members. The chapter additionally looks at safety considerations, effective operations, and operator training.

Chapter 8 by Flach and Bennett provides a focus that investigates coordination and control within distributed work settings with particular focus on the area of interface design. The chapter addresses the need for resilience in terms of flexible adaptation and provision of distributed authority to a number of autonomous agents. The chapter looks into design of interfaces that can yield this kind of authority while at the same time provide common ground to enable understanding of one's own actions in terms of the coordination with other agents who are working on dynamic problems. The chapter touches on the ideas central to polycentric control principles.

Chapter 9 by Coovert, Arbogast, and de Visser concludes the handbook with an appealing and pertinent area of research within distributed team cognition: developing artificial intelligence systems to work with distributed team members. The authors in particular address topics such as trust, humanness, morality, ethics, perception, and societal acceptance as they look this area through the construct of the "cognitive wingman." The chapter is a nice overview for looking at how interdisciplinary and joint aspects of artificial intelligence can be of benefit to distributed team cognition; deriving issues, principles, and benefits that need to be considered and come into play when systems are designed.

This concludes the three primers that outline the concept of each volume and what to expect within in terms of topics, direction, and content. The goal of this handbook is to take a broad, interdisciplinary perspective about distributed team cognition by providing representative samples of research, reviews, and selections that amplify our understanding in a comprehensive manner.

Michael D. McNeese

1 Wrapping Team Members' Heads around Managing Virtual Team-Related Paradoxes

John E. Mathieu, Jean Brittain Leslie, and Margaret M. Luciano

CONTENTS

Modern-day organizations use teams to align their human capital with organizational goals. But today's teams come in many different shapes and arrangements. Teams are formed, execute their actions, and disband at an alarming rate. Employees often work simultaneously in multiple teams with little cross-team coordination. Team memberships are fluid as individuals come and go so rapidly that it is often difficult to know who is on the team. And team members typically communicate and coordinate their efforts, at least in part, through virtual means. In short, the modern-day team landscape is complex and chaotic, and team members face many paradoxes as to how to work effectively in this context.

Our chapter considers how team members manage a number of paradoxes associated with operating in virtual team arrangements. Our primary thesis is that to the extent that members have a shared mental model of how they will handle such paradoxes their teams will be more effective. For purposes of this chapter, we adopt Kozlowski and Ilgen's (2006, p. 79) definition of work teams as:

> (a) two or more individuals who (b) socially interact (face-to-face or, increasingly, virtually); (c) possess one or more common goals; (d) are brought together to perform organizationally relevant tasks; (e) exhibit interdependencies with respect to workflow, goals, and outcomes; (f) have different roles and responsibilities; and (g) are together embedded in an encompassing organizational system, with boundaries and linkages to the broader system context and task environment.

The essence of this definition is that team members are interdependent and must plan and execute their actions to achieve common goals while operating within a particular context. Notably, team effectiveness is a multidimensional construct and can be gauged in terms of outcomes and by-products of team activity that are valued by one or more constituencies (Mathieu, Heffner, Goodwin, Salas, & Cannon-Bowers, 2000). Hackman (1990) identified three primary types of outcomes as: (a) performance, including quality and quantity, as evaluated by relevant others outside of the team; (b) meeting team member needs; and (c) viability, or the willingness of members to continue to work together as a team.

Given current communication and collaboration tools, team operating contexts are increasingly virtual these days. Maynard and Gilson (2014, p. 7) submitted that

> virtual teams have been defined as "functioning teams that rely on technology-mediated communication while crossing several different boundaries" (Martins, Gilson, & Maynard, 2004, p. 807) and a team is considered more or less virtual based on "the extent to which team members use virtual tools to coordinate and execute team processes" (Kirkman & Mathieu, 2005, p. 702).

A recent survey suggests that over two-thirds of multinational organizations utilize some form of virtual teaming (Society for Human Resource Management, 2012), which is only likely to grow in the future. Virtual teams (VTs) face many additional challenges by virtue of the fact that their members are dispersed across the globe in different time zones and often speak different languages. Therefore, better understanding how teams can coordinate their efforts in VTs represents an important challenge for organizational effectiveness.

Accordingly, this chapter explores how team members' shared mental models (SMMs) concerning work paradoxes relate to their effectiveness, especially when working primarily through virtual technologies. SMMs refer to "team members' shared understanding of team tasks, equipment, roles, goals, and abilities" (Lim & Klein, 2006, p. 403). We provide a brief review of how the nature (i.e., contents) of different mental models have been shown to be related to team processes and outcomes. We then submit that we should consider how members' SMMs concerning how to manage paradoxes relate to the effectiveness of virtual teams. In so doing we define paradoxes and highlight themes that are particularly salient for virtual teams. We demonstrate empirical relationships using a sample of 50 VTs. We close with recommendations concerning how various interventions could be leveraged to enhance virtual teams' SMMs concerning paradoxes.

SHARED MENTAL MODELS

Mental models are organized understandings or mental representations of knowledge (Cannon-Bowers, Salas, & Converse, 1993; Klimoski & Mohammed, 1994; Mohammed, Ferzandi, & Hamilton, 2010; Mohammed, Klimoski, & Rentsch, 2000). Sharedness of mental models is the extent to which team members' mental models are consistent with one another. SMMs represent an important team emergent state that enables members to coordinate their actions toward goal achievement. Marks, Mathieu, and Zaccaro (2001, p. 357) defined team emergent states as: "constructs that characterize properties of the team that are typically dynamic in nature and vary as a function of team context, inputs, processes, and outcomes" and offered SMMs as one prime example. Mathieu, Heffner, Goodwin, Cannon-Bowers, and Salas (2005) noted that SMM is a configural type of team construct and derives from the consistency of individuals' models—yet there is no "team model" per se. SMMs represent the extent to which team members' organize their knowledge structures in consistent ways so as to facilitate collective behavior. Cannon-Bowers et al. (1993) argued that teams can adapt quickly to changing task demands by drawing SMMs that enable them to predict what their teammates are going to do, and what they are going to need in order to do it. Thus, SMMs allow team members to determine and select different courses of action that are consistent and coordinated with those of their teammates.

Klimoski and Mohammed (1994, p. 432) suggested that "there can be (and probably would be) multiple mental models co-existing among team members at a given point in time." Mathieu et al. (2000) discussed numerous different types of mental models that they abstracted to two general types: task and team. Specifically, they suggested that *task-related SMMs* referred to members' understanding concerning the use of *technology or equipment*. Prior research has considered *task-related technologies* such as radar systems (air traffic controllers) or computer-aided design and manufacturing (CADCAM) systems, whereas virtual team applications may include the operation and use of collaborative tools (e.g., email, knowledge repositories, teleconferencing, document sharing). Alternatively, *team-related SMMs* refer to members' understanding as to how team interactions are orchestrated. These models describe members' SMMs concerning individual's roles and responsibilities, interaction patterns, decision-making processes, norms of behavior, and so forth. In short, how they will execute team processes (cf., Marks et al., 2001). The two forms of SMMs tend to overlap in VTs, however, as the use of collaborative and communication technologies coincides with determining roles and responsibilities, how, and when work will be accomplished.

Empirical findings have suggested members' SMMs relate significantly to their team processes (e.g., Marks, Sabella, Burke, & Zaccaro, 2002; Mathieu et al., 2000), other emergent states (e.g. Mathieu, Rapp, Maynard, & Mangos, 2009; Stout, Cannon-Bowers, & Salas, 2017), and team effectiveness (e.g., Ensley & Pearce, 2001; Kellermanns, Floyd, Pearson, & Spencer, 2008; Lim & Klein, 2006; Smith-Jentsch, Mathieu, & Kraiger, 2005). Indeed, based on meta-analytic findings of the SMM domain, DeChurch and Mesmer-Magnus (2010) concluded that when SMMs are indexed in terms of the consistency of members' knowledge structures, they are predictive of team processes and performance.

Despite the wealth of evidence that SMMs are advantageous for team effectiveness, Maynard and Gilson (2014, pp. 4–5) noted that

> To date, this work has assumed that all team member interaction is face-to-face. As such, there has been little attention given to how the use of information communication technologies (ICT) to communicate may affect the development of SMMs. We contend that not considering the effect of ICT usage on SMM development is an important omission because SMM development may be altered by the inherent attributes of the many ICT options currently available to teams.

Accordingly, we consider the particular challenges that teams who interact largely using ICT encounter, and how they relate to SMMs.

VIRTUAL TEAM PARADOXES

VTs are groups of geographically, organizationally, and/or time dispersed, mutually dependent workers brought together through technologies to work on the same objectives (Bell & Kozlowski, 2002). Their popularity is attributed to advancements in technology, globalization, and organizations' desires to be flexible, agile, and reduce operating costs. Effective virtual teams can benefit both employers and employees. Employers may use VTs to better leverage their human capital across time and locations. They may also benefit from reduced absenteeism, greater employee retention, and less overhead. Employees may benefit from flexibility, job satisfaction, and reduction in commuting time. However, not all virtual teams function effectively. Research shows managing virtual teams is more challenging than managing traditional face-to-face teams, and some experts suggest that more virtual teams fail than succeed (Gilson, Maynard, Jones Young, Vartiainen, & Hakonen, 2015). For instance, DeRosa (2010) lists six reasons that virtual teams fail: (1) ineffective leadership; (2) lack of clear goals, directions, or priorities; (3) lack of clear roles among team members; (4) lack of cooperation; (5) lack of engagement; and (6) inability to replicate a "high touch" environment. Elsewhere, Turmel (2018) cites five reason that virtual teams fail including: (1) teams lacking a compelling vision; (2) team members do not hold each other accountable for their work and deliverables; (3) the team doesn't have shared leadership; (4) team processes aren't effective or at least adhered to; and (5) problematic relationships with the manager.

The sources cited above, and scads of other commentaries, highlight the fact that critical challenges for VT effectiveness include leadership, communication, integrating perspectives, and coordinating action. While often viewed as problems to be resolved, choices to be made, or balances to be struck, these themes might be better conceived of as paradoxes to manage. Paradoxes "denote persistent contradictions between interdependent elements. While seemingly distinct and oppositional, these elements actually inform and define one another, tied in a web of eternal mutuality" (Schad, Lewis, Raisch, & Smith, 2016, p. 6). The interdependence between the elements or pairs creates both a tension and an opportunity. Virtual teams, for example, need to both plan and execute, should be both diverse in their thinking and unified in their actions, and must focus both on short-term and long-term goals. Paradoxes, as defined here, show up in all facets of organizational life. They are not problems that can be easily solved with one-time solutions; rather, they are of cyclical or reoccurring nature. The more strongly people become attached to one side of a pair, the harder it is to "see" its negatives (Johnson, 2014).

Helping VTs deal with challenges may sometimes be as easy as helping them decide which solution is most appropriate for their circumstances (e.g., which means of collaboration is best suited for a given function—such as email or a threaded discussion list). But many challenges require more sophisticated approaches. In his acclaimed book *Opposable Mind*, Roger Martin (2009) points out a common theme from his interviews with business leaders, which is that successful leaders "have the predisposition and the capacity to hold two diametrically opposing ideas in their heads" (p. 6) and they are "able to produce a synthesis that is superior to either opposing idea" (p. 6). He goes on, "the ability to use the opposable mind is an advantage at any time, in any era" (p. 8). A well-managed polarity or paradox occurs when teams capitalize on the inherent tensions between the interdependent pairs while avoiding the limits of either. In other words, when virtual teams can see a paradox as two equally important points of view, they can begin to exploit the benefits inherent in the pairs.

Based on the extant literature and qualitative grounding with VT leaders and members, we identified numerous challenges that are best conceived of as paradoxes. For illustrative purposes we feature four of these in Table 1.1 and in the discussion that follows. Each paradox consists of two interdependent poles or themes. Each theme has the potential to benefit or hinder (if focused on to the neglect of its partner theme) VT effectiveness.

Leadership—Task and Relationship

Virtual teams require effective leadership to help establish high quality working relationships. The formation of effective working relationships often involves numerous formal (e.g., team building events) and informal (e.g., having lunch or coffee together, chatting by the water cooler) interactions, which are costly if not impossible to duplicate in virtual teams. Interpersonal challenges may arise for a number of reasons, including a lack of accountability, a lack of attendance or engagement in team-building activities, and focusing on non-task issues. An underlying paradox to this challenge is the team being attentive to both *task* and *relationship* leadership behaviors. Task-oriented leadership behaviors include establishing shared norms, negotiation, and holding the team accountable for its performance and outcomes.

TABLE 1.1
Virtual Team Paradoxes

Paradox	Paradox Description	Example
Leadership—task focused and relationship oriented	Virtual teams must be task focused to be effective. At the same time, they must also focus on building relationships across time, culture, and distance to succeed as a team.	"We tend to probably be very task focused working to just get things done on time and I think where we probably suffer a bit is to take advantage of each other."
Communication—formal and informal	Virtual teams rely heavily on effective formal communication. At the same time, successful VTs leverage the value of informal communication.	"We spend time daily chatting back and forth via telephone, via IM, and via face-to-face for folks who are in the same site. We have weekly and biweekly staff meetings."
Perspective—unified team and diverse individuals	Virtual team members must act as one unified team. At the same time, they must maintain their distinct perspectives and identities.	"Work is primarily individual ('we are the masters of our own success'), they each do the same procedure and have their own targets. Each person works in a clearly defined geographical scope, but there are certain topics on which the team members work together."
Synchronicity—working apart and together	Virtual teams work toward common goals while being geographically dispersed. At the same time, they need face-to-face time to bond as a team and accomplish complex tasks.	"When we are together (in the same room) it's easier to brainstorm out loud, our decision making is best made when we are all together, and that's our time to check in and make sure we are all on the same page."

Task-oriented behaviors are critical to assure the work of the team is delivered on time and that there is a sense of progress and pride in the team's work. Relationship-oriented leadership behaviors include attending to members' well-being, nurturing team identity, maintaining a sense of inclusion, and promoting positive relationships. Managing team social and interpersonal interactions is a critical team process that lays the foundation for the effectiveness of other processes. Overemphasis on the task to the neglect of relationships can result in team members failing to form a cohesive team and reduction in helping behaviors, whereas overemphasis on relationships to the neglect of the task can result in missed deadlines and losing sight of the team's objectives.

Communication—Formal and Informal

Effective communication is an important aspect of virtual teamwork. Communication issues may arise for a number of reasons, including the failure of members to comprehend that content the other members intend, questions not being answered correctly and/or not being directed to the right person, the failure to distribute information to all team members, problems using communication media, difficulty in conveying and understanding the importance of certain information, time delays, misinterpretations of silence, differing interests or goals, anxiety or uncertainty, and cultural barriers. For virtual teams to be effective, it is necessary to attend to both *formal* (e.g., memos, meetings) and *informal communication* (e.g., IM, GTalk). Formal communication is useful to share large amounts of important information and to establish a shared understanding of team responsibilities and goals. Informal communication helps team members to get real-time answers, deepen relationships, and align individual perspectives to team goals. Strict use of formal communication can lead to time-consuming meetings and delays while waiting for formal responses. Whereas strict use of informal communication can lead to team members being unclear about team goals and responsibilities and taking inappropriate actions.

Perspective—Unified Team and Diverse Individuals

VTs are assembled to bring together individuals with unique competencies, experiences, and perspectives; however, those differences may result in miscommunications or team members pursuing different directions and failing to come together as a cohesive team. The underlying paradox to this challenge is a strong team requires a duality of foci and integrative thinking that include *diverse individuals* that offer their own unique perspectives and a *unified team* that agrees on how to move forward. Both diverse member perspectives and a unified team perspective have potential benefits and detriments. Diverse perspectives can be a source of strength and innovation, yet overemphasized, the team members may be working in different directions or embroiled in dysfunctional conflict. Similarly, a unified team viewpoint promotes unified action towards a common purpose, yet overemphasized, may result in groupthink and stagnation. The perspective integration paradox challenges teams to find unity in the differences.

Synchronicity—Working Apart and Together

Coordinating and combining efforts are important aspects of virtual teamwork. For many virtual teams, being all physically together is not an option or an infrequent option, so determining when and how to work together (physically or virtually) becomes more difficult and more vital. Challenges coordinating work efforts may arise for a number of reasons, including members who are globally dispersed, overemphasize a traditional schedule, or have insufficient technology. For virtual teams to be effective, it is necessary to work both *apart* and *together*—both asynchronously and synchronously. Working apart is useful to allow each team member to focus on his/her individual tasks and contributions to problem solving are richer because each team member has access to local resources. Working together is useful to solve key issues, develop richer relationships, and dedicate focused attention to a particular team task. If teams overemphasize working apart, solving issues independently can result in conflicting ideas, solutions, and products that lack integrations. If teams overemphasize working together, progress may slow as too many people are working on the same task and problem solving may be hindered by a lack of alternative perspectives.

In sum, VTs face many challenges in terms of how to best manage paradoxes associated with leadership, communication, perspective, and synchronicity. We submit that there is not necessarily one optimal way to approach such challenges, but rather, what matters is the extent to which team members have SMMs in terms of how paradoxes should be handled. We believe that team members' degree of SMMs on the four paradoxes identified above will related positively to team effectiveness in terms of their performance, viability, and member reactions.

METHOD

SAMPLE

As part of a larger investigation, over 140 VTs were recruited from 56 for-profit, non-profit, and government organizations from a wide variety of industries and geographic regions. Participating companies were recruited through personal contacts of the authors, postings on virtual team discussion groups, and clients who had previously participated in programs at the Center for Creative Leadership. The larger investigation was rolled out over 18 months and surveyed VT members and their leaders twice approximately six months apart. Some of the teams participated in interventions designed to raise awareness of paradoxes and how they might be managed, whereas other teams served as quasi-experimental controls. For purposes of the present investigation, we report findings from the initial phase from 307 individuals from 50 teams from 31 organizations for whom we had 4 to 12 member survey responses, and ratings of their performance from an external leader. None of the teams had participated in the interventions when these data were collected.

The sample average age was 42.6 (SD = 10) with an average team tenure of 2.8 years (SD = 3.0). Forty-eight percent of the sample was male and 52% female. The sample came from 18 different time zones, 35 different countries with the highest proportions from the United States (45%), China (11%), and India (5%), and was generally well-educated with 10.4% having doctoral level degrees, 38.1% master's level degrees, and 40.7% bachelor's degrees.

MEASURES

Members completed an online survey that included measures the four paradoxes, their work-related reactions, and features of their VT context. Team leaders completed an online survey that included measures of their demographics and team performance.

Team Virtuality

Following Kirkman and colleagues' conceptualizations of virtuality (cf., Kirkman & Mathieu, 2005; Kirkman, Rosen, Tesluk, & Gibson, 2004), we assessed the extent to which team members used various information and communication technological tools. Specifically, we asked team members to indicate the percentage of their VT-related time spent: (1) completing individual work (M = 36%); (2) meeting face-to-face (M = 13%); (3) conducting conference calls (M = 15%); (4) exchanging emails (M = 17%); (5) video conferencing (M = 4%); (6) sharing documents (M = 4%); (7) exchanging instant messaging/texting (M = 4%); or (8) using other technologies (M = 7%). On average, teams reported using virtual tools for slightly over half of their interactions (M = 50.6%, SD = 25%). Moreover, team members evidenced significant interrater reliability [$F(49, 257) = 2.38$, $p < .001$; ICC1 = .18; ICC2 = .58] concerning their use of virtual tools. ICC1 indexes the reliability of individual ratings of the group construct, whereas ICC2 represents the reliability of the group average rating. Accordingly, we averaged team member's ratings to yield a measure of team virtuality.

Team Paradox SMMs

We assessed four paradoxes, each using 12-item measures from Leslie, McCauley, McPartlan, and Barts (2014). As illustrated in Table 1.2, each of the four paradoxes had two alternatives or themes. Respondents were asked to indicate how often each statement was true for their team using the following five-point scale: (1) almost never, (2) seldom, (3) somtimes, (4) often, and (5) almost always. For example, the leadership paradox included a task focused and relationship focus pair of themes. For each paradox, three positively worded and three negatively worded items were presented for each theme, yielding a total of 12 items. Note that the seemingly alternative pairings, as well as the combination of positively and negatively worded items per theme, generated sufficient variance in member responses to calculate inter-rater reliability. As is convention in SMM literature (cf., Mohammed et al., 2000; Smith-Jentsch et al., 2005), we correlated each member's ratings with each other teammate's ratings across the 12 items per paradox. We then averaged those inter-member correlations to yield an overall team-level average SMM index. Higher average correlations represent greater consistency or similarity of mental models. Example items for each paradox are presented in Table 1.2.

Team Viability

Team viability refers to the extent that the team is likely to remain together in the future and was measured using two items: (1) I wouldn't hesitate to participate on another task with the same team members; and (2) If given a choice, I prefer to work with another team rather than this one (reverse coded). Respondents used a seven-point agreement scale for both items: (1) strongly disagree, (2) disagree, (3) slightly disagree, (4) neutral, (5) slightly agree, (6) agree, and (7) strongly agree. Responses to the two items correlated significantly [$r = .27$, $p < .001$] so we averaged them to yield a measure of team viability. Team members evidenced significant interrater reliability [$F(49, 257) = 2.36$, $p < .001$; ICC1 = .18; ICC2 = .58] concerning their team's viability, so we averaged their responses per team.

Individual Reactions

Team members responded to five items concerning their personal reactions to their VT participation using the seven-point agreement scale detailed above. They indicated their team trust and commitment using the following two items: "I *trust* this team" and "I am *committed* to this team," respectively. They also indicated their

TABLE 1.2
Paradox Themes and Example Items

Paradox	Themes and Items
Leadership	**Task Focused**
	1. Team members are held accountable for their performance and outcomes.
	2. Team members focus on finishing tasks rather than on building relationships.
	Relationship Oriented
	1. The team's culture supports positive relationships.
	2. The team spends so much time on building relationships that deadlines are missed.
Communication	**Formal**
	1. Formal team meetings help to build the team.
	2. The team stalls because of its need for formal agreement or consensus.
	Informal
	1. Team members use informal conversation and connections to get real-time answers for making quick progress on their work.
	2. Too much informal communication results in cliques that create rifts between team members.
Perspective	**Unified Team**
	1. There is agreement among team members about what the team needs to accomplish.
	2. There is reluctance to suggest new ideas that the team might reject.
	Diverse Individual
	1. Diverse perspectives within the team stimulate new ideas.
	2. Members are unable to agree on what the team needs to accomplish.
Synchronicity	**Working Apart**
	1. Contributions to problem solving are richer because each team member has access to local resources.
	2. The team's work can lack integration when members work separately.
	Working Together
	1. The team leverages the energy from being together to solve key issues.
	2. Conflict is amplified when the team spends too much time together.

Note: Items numbered "2" are reversed worded for their respective paradox theme

personal development attributable to working on their VT using the following three items: (1) being a member of this team contributes to my own learning and development; (2) working on this team has provided me with the opportunity for professional growth and development; and (3) I have learned a lot of valuable work-related information by being a member of this team. The three items evidence a scale reliability of $\alpha = .86$ so we averaged them as a measure of *member development*.

Team Performance

Team leaders rated the performance of their VT using a five-item measure ($\alpha = .88$) adapted from Maynard, Mathieu, Rapp, and Gilson (2012). Sample items include

"This team achieves its goals" and "This team does high-quality work." Leaders rated each item using a seven-point response scale that ranged from (1) very inaccurate, (2) mostly inaccurate, (3) slightly inaccurate, (4) uncertain, (5) slightly accurate, (6) mostly accurate, and (7) very accurate. We averaged the five items as an index of team performance.

RESULTS

We discuss the findings from this investigation at two levels of analysis. First, at the team level of analysis, we test whether team members' SMMs of paradox management relate significantly to team performance and viability using regression analyses. Second, we test cross-level models that relate team-level paradox SMMs with individual team member reactions using a hierarchical linear modeling (HLM). HLM controls for the fact that team members are non-independent by virtue of being members of the same team (Raudenbush, Bryk, Cheong, Congdon, & du Toit, 2004).

TEAM LEVEL

The correlations between team virtuality, the four paradox SMMs, and team performance and viability are presented in Table 1.3. As shown, all four paradox SMMs correlated significantly with both team performance [rs = .31 to .50, $p < .05$] and team viability [rs = .44 to .65, $p < .05$]. We then regressed team performance onto team virtuality [$\beta = .11$, ns] and the four SMMs and found that only synchronicity [$\beta = .39$, $p < .05$] exhibited a significant unique relationship [leadership: $\beta = .33$, ns; communication: $\beta = -.26$, ns; perspective: $\beta = .12$, ns]. Collectively, the SMMs accounted for 33% of the variance of team performance. No doubt the high intercorrelations among the SMMs limited the ability to discern unique contributions.

TABLE 1.3
Team-Level Correlations

Variables	1	2	3	4	5	6	7	8
1. Virtuality	—							
SMMs								
2. Leadership	–.23	—						
3. Communication	–.18	.72	—					
4. Perspective	–.06	.74	.62	—				
5. Synchronicity	–.26	.69	.67	.60	—			
6. SMM4[a]	–.21	.90	.88	.85	.85	—		
7. Performance	–.07	.49	.31	.44	.50	.49	—	
8. Viability	–.11	.64	.55	.44	.65	.65	.52	
Mean	50.7	.65	.50	.61	.40	.54	5.9	5.9
SD	15.1	.20	.24	.22	.20	.19	.61	.51

Note: N = 50 teams; correlations > |.28|, $p < .05$; > |.36|, $p < .01$

[a] Average of the four SMMs

Indeed, regressing performance onto virtuality and an SMM composite (derived by averaging the four paradoxes) revealed a positive relationship for SMMs taken as a whole [$\beta = .50$, $p < .001$].

We next regressed team viability onto team virtuality [$\beta = .06$, ns] and the four SMMs, which revealed significant unique relationships for synchronicity [$\beta = .42$, $p < .05$] and leadership [$\beta = .47$, $p < .05$], but not for communication [$\beta = .06$, ns] or perspective [$\beta = -.19$, ns]. Collectively, the SMMs accounted for 51% of the variance of team viability. Here again, regressing team viability on to virtuality and the SMM composite revealed a positive relationship for SMMs taken as a whole [$\beta = .65$, $p < .001$].

In sum, at the team level of analysis, members' paradox SMMs uniformly correlated positively with both team performance and viability. Synchronicity exhibited unique significant positive relationships in both regressions, and leadership did in the viability equation. However, it appears as though there may be a "gestalt" type relationship, as an average of the four SMM measures contributed significantly to the prediction of both team-level outcomes beyond that accounted for by team virtuality and synchronicity alone. Getting members on the same page clearly benefits team-level outcomes.

Cross Level

The correlations between team virtuality, the four paradox SMMs, and the individual level outcomes are presented in Table 1.4. As shown, all four paradox SMMs correlated significantly with each of the member reactions [rs = .16 to .41, $p < .05$]

TABLE 1.4
Individual and Cross-Level Correlations

Variables [a]	1	2	3	4	5	6	7	8	9
1. Virtuality	—								
SMMs									
2. Leadership	–.22	—							
3. Communication	–.15	.74	—						
4. Perspective	–.06	.76	.68	—					
5. Synchronicity	–.18	.70	.71	.62	—				
6. SMM4[b]	–.17	.90	.90	.87	.85	—			
Member Reactions									
7. Trust	–.00	.41	.36	.37	.33	.42	—		
8. Commitment	–.05	.22	.16	.20	.20	.22	.54	—	
9. Member development	–.02	.40	.36	.39	.29	.41	.56	.53	—
Mean	50.7	.65	.48	.61	.40	.54	6.1	6.5	6.2
SD	13.7	.20	.24	.22	.20	.19	1.01	.69	.90

Note: N = 307 members in 50 teams; correlations > |.11|, $p < .05$; > |.15|, $p < .01$

[a] Team variables assigned to individuals, so significance levels should be interpreted cautiously

[b] Average of the four paradox SMMs

although the significance values should be interpreted cautiously as they have not been adjusted for nonindependence. Indeed, each of the three member reactions exhibited significant ($p < .001$) between team variance warranting the use of HLM for testing substantive relations [i.e., trust = 23%, commitment = 6%, member development = 22%].

The HLM results are summarized in Table 1.5. For each criterion, we first controlled for members' individual differences and then introduced team virtuality and the four SMMs to the equation. As shown, none of the individual differences evidenced any significant effects save for a positive relationship between members' age and their team commitment [$\beta = .13$, SE = .06, $p < .05$]. Team virtuality did not relate significantly to any of the member reactions, and few of the SMMs evidenced any significant unique relationships. The leadership SMM did relate significantly with members' trust [$\gamma = .28$, SE = .09, $p < .001$] and development [$\gamma = .24$, SE = .09, $p < .001$], and perspective also related significantly to member development [$\gamma = .16$, SE = .08, $p < .05$], but no other unique effects were evident. As with the team-level analyses, however, the likely culprit is high correlations among the SMMs measures. We recalculated the analyses chronicled in Table 1.5 substituting the composite SMM measure (i.e., average) for the four specific ones and obtained a significant positive relationship in each equation [trust: $\gamma = .43$, SE = .07, $p < .001$; commitment: $\gamma = .22$, SE = .07, $p < .01$; and member development: $\gamma = .40$, SE = .05, $p < .001$].

TABLE 1.5
Cross-Level Relationships between Paradox SMMs and Member Reactions

Criteria

Predictors	Trust	Commitment	Member Development
Covariates			
1. Sex [a]	.06 (.05)	–.06 (.04)	–.01 (.05)
2. Age	.06 (.06)	.13 (.06)*	–.02 (.07)
3. Education	.03 (.05)	.08 (.06)	–.04 (.06)
4. Team tenure	.06 (.04)	.05 (.08)	–.02 (.06)
Cross-Level Effects			
5. Virtuality	.08 (.05)	–.05 (.07)	.06 (.05)
SMMs			
6. Leadership	.28 (.09)**	.13 (.13)	.24 (.09)**
7. Communication	.11 (.08)	–.02 (.10)	.09 (.08)
8. Perspective	.09 (.09)	.09 (.11)	.16 (.08)*
9. Synchronicity	–.00 (.10)	.04 (.09)	–.03 (.09)
~R^2	.23	.06	.22

Note: Table values are HLM parameter estimates, standard errors within parentheses.
N = 307 members in 50 teams
* p < .05
** p < .01
[a] Coded: women = 0, men = 1

In sum, the cross-level effects suggest that the extent to which team members have SMMs concerning the paradoxes relates significantly to their reactions to working in VT environments. Of the four paradoxes, SMMs concerning leadership appear to be the most potent, contributing uniquely to the prediction of both member trust and development. Yet the general theme appears that having high SMMs concerning the four paradoxes contributes positively to member trust in, and commitment to, their VT, and the extent to which they report development associated with the experience.

DISCUSSION

The literature to date has shown that members' SMMs of task and team properties are associated positively with team effectiveness. We have extended that discussion to consider SMMs of VT paradoxes. A key consideration of our work is that we do not model the "accuracy" of members SMMs per se—as paradoxes defy conventional logic in terms of accurate and inaccurate perceptions. Paradoxes are often best addressed as "both-and" rather than "either-or" propositions, as members seek to manage them rather than to resolve them. In those contexts, then, what appears important is that members are in concert with one another about how they are managing these paradoxes. Our findings suggest that to the extent that members' have overall SMMs, all three facets of team effectiveness—i.e., performance, viability, and member reactions—are enhanced. More specifically, of the four paradoxes, at the team level of analysis the synchronicity SMM exhibited positive unique relationships with both performance and viability, whereas in terms of cross-level effects, the leadership SMM had significant unique positive relationships with member trust and development. Yet the general pattern appeared to be that all facets of team effectiveness were related positively to overall SMMs considered across the four paradoxes, suggesting the potential power of a gestalt type relationship.

Of course, the findings reported herein are not definitive in terms of causal inferences as they were found in the context of a cross-sectional design. Stronger research designs where both SMMs and team effectiveness measures are gathered and modeled longitudinally would be preferable. Moreover, introducing various interventions would lend more credence to causal interpretations. With that caveat in mind, presuming that these relationships do reveal potential leverage points, the question becomes how to promote SMMs regarding these VT paradoxes for the benefit of teams and members alike.

Whereas these findings are encouraging, they do raise a number of related questions for scholars and practitioners alike. For example, to the extent that VTs include members from around the globe (i.e., GVTs), cultural differences may be important variables to consider. For instance, Mohammed, Hamilton, Tesler, Mancuso, and McNeese (2015) advanced the idea that SMMs in terms of temporal issues are an important ingredient for team success. We know that cultures vary widely in their conceptions and approach to time (e.g., Briley, 2013), which could serve to challenge the coordination of GVTs.

Furthermore, employees who work in VTs rarely are members of a single team (O'Leary, Mortensen, & Woolley, 2011). Maynard et al. (2012) found that members' average percentage of time working on a focal team (i.e., fewer other team membership demands) related positively to the planning processes of highly interdependent teams and thereby to their transactive memory systems and effectiveness. Multiple-team memberships create pressures on employees in terms of their motivations, identity, and task switching. For instance, Rapp and Mathieu (2019) found that the variety of teams that an employee works on simultaneously and the associated role stress related negatively to their identification with any given team. Moreover, facets of those team memberships such as their relative cohesion, prestige, and stage of completion also impacted employees' identifications with their different memberships. No doubt the diffusion of employees' attention and other switching costs (e.g., Altmann & Gray, 2008) associated with attempting to work on multiple teams simultaneously needs to be gauged against the potential benefits of deploying members to multiple teams.

Finally, the role of new and emerging types of collaboration tools and their implications for team coordination need to be considered. For instance, what are the implications of simultaneous document editing for the coordination of knowledge focused teams? What are the implications of smartphone-enabled, 24/7 availability on members' coordination and stress levels, especially for GVTs with members in markedly different time zones? Paradoxes come in many different forms, and new and emerging work arrangements will likely exacerbate their challenges.

Applied Implications

The SMM literature has been mostly correlational with relatively few interventions investigated. One often recommended intervention focuses on training (e.g., Cannon-Bowers, Salas, & Converse, 1990; Stout et al., 2017). For instance, Marks, Zaccaro, and Mathieu (2000) illustrated how team interaction training could enhance members' SMMs, whereas Marks et al. (2002) illustrated how cross-training members could also enhance SMMs. Yet, while there is evidence that task or team SMMs can be enhanced through training, it is not clear whether paradox SMMs can be influenced in a similar way. There is a clear need for future investigations that test whether training interventions can influence how team members mange paradoxes, especially in field settings.

There are other interventions that might be considered for enhancing VT paradox SMMs. For instance, Edwards, Day, Arthur, and Bell (2006) suggested that team composition configurations could be orchestrated to promote SMMs. However, the very nature of VTs is to bring together diverse members who can contribute to the task at hand. Perhaps an alternative to compositional approaches might be to introduce early interventions that promote paradox thinking at the beginning of a VT team life cycle. For instance, Mathieu and Rapp (2009) illustrated how team charters and formal planning helped to set teams on the right trajectory and to sustain their high performance over time. Haig, Sutton, and Whittington (2006) describe how situation, background, assessment, recommendation (SBAR) training can facilitate the formation of SMMs in healthcare. Elsewhere, Marks et al. (2000) illustrated how initial leader briefings could be used to promote SMMs. In short, initial interventions that catch teams early in their life cycles could well prove to be beneficial by helping to forge SMMs concerning how they will approach and handle VT paradoxes.

Both logistical and substantive factors may limit the extent to which VTs can "get off on the right foot" initially. VTs are often formed very quickly and with people from across the globe. It may be more often the norm than exception that there is little time for foundational interventions. Moreover, it may be difficult to discuss paradoxes before teams have actually launched and begun their work. Paradox thinking is very abstract and addressing how to best manage them may be enhanced once members have some experience working with one another and have confronted the issues. In other words, some time after launch, but before teams have cemented their interaction processes, may prove to be an ideal time to address team paradox SMMs. As a base for this team training, organizations may consider individual training on how to engage paradoxical thinking (holding and integrating opposing ideas), which could be done asynchronously.

Other potential interventions that can be leveraged after teams have begun working may come in different forms. First, the concept of team reflexivity describes team members' inclination to naturally review their previous interactions and to derive insights and lessons learned. Gurtner, Tschan, Semmer, and Nägele (2007) illustrate how reflexivity can be prompted by different interventions. In particular, Tannenbaum and Cerasoli (2013) reported meta-analytic results of the benefits of after-action reviews (AARs)—or team debriefs—on later team processes and effectiveness. We propose that paradox oriented AARs/debriefs could be applied to promote SMMs.

Conclusion

VTs are prominent in today's work environment and are only likely to be more so in the years to come. VTs enable organizations to align and redeploy their human capital quickly and efficiently to meet organization demands. Yet the benefits of VT arrangements bring with them many challenges—hence the paradox of how to best manage them to maximize their effectiveness. We suggested that to the extent that VT team members have SMMs as to how they will manage paradoxes, their team performance and viability will be enhanced, as will their personal reactions concerning working in VT environments. We provided empirical results that were consistent with this thesis using data from 50 different VTs. Four different team-related paradox SMMs evidenced significant correlations with team- and member-level effectiveness criteria, and additional analyses suggest that the SMMs operated in concert with one another such that an overall composite illustrated universally positive relationships with the effectiveness criteria. We conclude that helping VTs to better manage their inherent paradoxes offers a potentially valuable intervention.

ACKNOWLEDGMENT

This study was funded by a grant from the SHRM Foundation (#166). However, the interpretations, conclusions, and recommendations are those of the author(s) and do not necessarily represent the views of the SHRM Foundation.

REFERENCES

Altmann, E. M., & Gray, W. D. (2008). An integrated model of cognitive control in task switching. *Psychological Review, 115*, 602–639. https://doi.org/10.1037/0033-295X.115.3.602

Bell, B. S., & Kozlowski, S. W. (2002). A typology of virtual teams: Implications for effective leadership. *Group & Organization Management, 27*(1), 14–49.

Briley, D. A. (2013). Looking forward, looking back: Cultural differences and similarities in time orientation. In *Understanding culture* (pp. 315–329). New York: Psychology Press.

Cannon-Bowers, J. A., Salas, E., & Converse, S. (1990). Cognitive psychology and team training: Training shared mental models and complex systems. *Human Factors Society Bulletin, 33*(12), 1–4.

Cannon-Bowers, J. A., Salas, E., & Converse, S. A. (1993). Shared mental models in expert team decision making. In J. N. J. Castellan (Ed.), *Current issues in individual and group decision making* (pp. 221–246). Hillsdale, NJ: Erlbaum.

DeChurch, L. A., & Mesmer-Magnus, J. R. (2010). Measuring shared team mental models: A meta-analysis. *Group Dynamics: Theory, Research, and Practice, 14*(1), 1–14.

DeRosa, D. (2010). *6 common reasons virtual teams fail.* Leadership Development Blog.

Edwards, B. D., Day, E. A., Arthur Jr, W., & Bell, S. T. (2006). Relationships among team ability composition, team mental models, and team performance. *Journal of Applied Psychology, 91*(3), 727.

Ensley, M. D., & Pearce, C. L. (2001). Shared cognition in top management teams: Implications for new venture performance. *Journal of Organizational Behavior, 22*(2), 145–160.

Gilson, L. L., Maynard, M. T., Jones Young, N. C., Vartiainen, M., & Hakonen, M. (2015). Virtual teams research: 10 years, 10 themes, and 10 opportunities. *Journal of Management, 41*(5), 1313–1337.

Gurtner, A., Tschan, F., Semmer, N. K., & Nägele, C. (2007). Getting groups to develop good strategies: Effects of reflexivity interventions on team process, team performance, and shared mental models. *Organizational Behavior and Human Decision Processes, 102*(2), 127–142.

Hackman, J. R. (1990). Work teams in organizations: An orienting framework. In J. R. Hackman (Ed.), *Groups that work (and those that don't)* (pp. 1–14). San Francisco, CA: Jossey-Bass.

Haig, K. M., Sutton, S., & Whittington, J. (2006). SBAR: A shared mental model for improving communication between clinicians. *The Joint Commission Journal on Quality and Patient Safety, 32*(3), 167–175.

Johnson, B. (2014). Reflections: A perspective on paradox and its application to modern management. *The Journal of Applied Behavioral Science, 50*(2), 206–212.

Kellermanns, F. W., Floyd, S. W., Pearson, A. W., & Spencer, B. (2008). The contingent effect of constructive confrontation on the relationship between shared mental models and decision quality. *Journal of Organizational Behavior, 29*(1), 119–137.

Kirkman, B. L., & Mathieu, J. E. (2005). The dimensions and antecedents of team virtuality. *Journal of Management, 31*(5), 700–718.

Kirkman, B. L., Rosen, B., Tesluk, P. E., & Gibson, C. B. (2004). The impact of team empowerment on virtual team performance: The moderating role of face-to-face interaction. *Academy of Management Journal, 47*(2), 175–192.

Klimoski, R., & Mohammed, S. (1994). Team mental model: Construct or metaphor? *Journal of Management, 20*(2), 403–437.

Kozlowski, S. W., & Ilgen, D. R. (2006). Enhancing the effectiveness of work groups and teams. *Psychological Science in the Public Interest, 7*(3), 77–124.

Leslie, J. B., McCauley, C. D., McPartlan, P. B, & Barts, D. (2014). *Virtual teams polarity assessment* [Polarity Assessment™ Software]. Unpublished instrument.

Lim, B. C., & Klein, K. J. (2006). Team mental models and team performance: A field study of the effects of team mental model similarity and accuracy. *Journal of Organizational Behavior, 27*(4), 403–418.

Management, S. H. R. (2012). SHRM: Virtual teams utilized by nearly half of organizations surveyed. Retrieved from https://www.shrm.org/about-shrm/press-room/press-releases/pages/virtualteamspoll.aspx

Marks, M. A., Mathieu, J. E., & Zaccaro, S. J. (2001). A temporally based framework and taxonomy of team processes. *Academy of Management Review, 26*(3), 356–376.

Marks, M. A., Sabella, M. J., Burke, C. S., & Zaccaro, S. J. (2002). The impact of cross-training on team effectiveness. *Journal of Applied Psychology, 87*(1), 3–13.

Marks, M. A., Zaccaro, S. J., & Mathieu, J. E. (2000). Performance implications of leader briefings and team-interaction training for team adaptation to novel environments. *Journal of Applied Psychology, 85*(6), 971–986.

Martin, R. L. (2009). *The opposable mind: How successful leaders win through integrative thinking*. Boston, MA: Harvard Business School Press.

Martins, L. L., Gilson, L. L., & Maynard, M. T. (2004). Virtual teams: What do we know and where do we go from here? *Journal of Management, 30*, 805–835.

Mathieu, J. E., Heffner, T. S., Goodwin, G. F., Cannon-Bowers, J. A., & Salas, E. (2005). Scaling the quality of teammates' mental models: Equifinality and normative comparisons. *Journal of Organizational Behavior, 26*(1), 37–56.

Mathieu, J. E., Heffner, T. S., Goodwin, G. F., Salas, E., & Cannon-Bowers, J. A. (2000). The influence of shared mental models on team process and performance. *Journal of Applied Psychology, 85*(2), 273–283.

Mathieu, J. E., & Rapp, T. L. (2009). Laying the foundation for successful team performance trajectories: The roles of team charters and performance strategies. *Journal of Applied Psychology, 94*(1), 90–103.

Mathieu, J. E., Rapp, T. L., Maynard, M. T., & Mangos, P. M. (2009). Interactive effects of team and task shared mental models as related to air traffic controllers' collective efficacy and effectiveness. *Human Performance, 23*(1), 22–40.

Maynard, M. T., & Gilson, L. L. (2014). The role of shared mental model development in understanding virtual team effectiveness. *Group & Organization Management, 39*(1), 3–32.

Maynard, M. T., Mathieu, J. E., Rapp, T. L., & Gilson, L. L. (2012). Something (s) old and something (s) new: Modeling drivers of global virtual team effectiveness. *Journal of Organizational Behavior, 33*(3), 342–365.

Mohammed, S., Ferzandi, L., & Hamilton, K. (2010). Metaphor no more: A 15-year review of the team mental model construct. *Journal of Management, 36*(4), 876–910.

Mohammed, S., Hamilton, K., Tesler, R., Mancuso, V., & McNeese, M. D. (2015). Time for temporal team mental models: Expanding beyond "what" and "how" to incorporate "when". *European Journal of Work and Organizational Psychology, 24*(5), 693–709.

Mohammed, S., Klimoski, R., & Rentsch, J. R. (2000). The measurement of team mental models: We have no shared schema. *Organizational Research Methods, 3*, 123–165.

O'Leary, M. B., Mortensen, M., & Woolley, A. W. (2011). Multiple team membership: A theoretical model of its effects on productivity and learning for individuals and teams. *Academy of Management Review, 36*(3), 461–478.

Rapp, T. L., & Mathieu, J. E. (2019). Team and individual influences on members' identification and performance per membership in multiple team membership arrangements. *Journal of Applied Psychology 104*(3), 303–320.

Raudenbush, S. W., Bryk, A. S., Cheong, Y. F., Congdon, R. T., & du Toit, M. (2004). *HLM6: Hierarchical linear and nonlinear modeling*. Lincolnwood, IL: Scientific Software International, Inc.

Schad, J., Lewis, M. W., Raisch, S., & Smith, W. K. (2016). Paradox research in management science: Looking back to move forward. *Academy of Management Annals*, *10*(1), 5–64.

Smith-Jentsch, K. A., Mathieu, J. E., & Kraiger, K. (2005). Investigating linear and interactive effects of shared mental models on safety and efficiency in a field setting. *Journal of Applied Psychology*, *90*(3), 523–535.

Stout, R. J., Cannon-Bowers, J. A., & Salas, E. (2017). The role of shared mental models in developing team situational awareness: Implications for training. In *Situational awareness* (pp. 287–318). New York: Routledge.

Tannenbaum, S. I., & Cerasoli, C. P. (2013). Do team and individual debriefs enhance performance? A meta-analysis. *Human Factors*, *55*(1), 231–245.

Turmel, W. (2018). Five reasons remote teams fail. In *The connected manager—Better management in a virtual world*. Retrieved from https://www.management-issues.com/connected/7159/five-reasons-remote-teams-fail/

2 Stabilizing Digital Infrastructures in Distributed Social Science Collaboration

Steve Sawyer, Jaime Snyder, Matt Willis, Sarika Sharma, Carsten Østerlund, and Emma Allen

CONTENTS

We focus on the uses of digital resources by social scientists collaborating at a distance in order to pursue insights into "infrastructure in use." In doing so we advance the concept of stabilization as a sociotechnical activity that demands extensive negotiations relative to the structure, form, uses, and arrangements of digital materials. And, we situate the work of stabilization relative to the extensive literature and developing conceptual basis of distributed cognition (Hutchins, 1995; Salas & Fiore, 2004; Fiore & Wiltshire, 2016; Nemeth, O'Connor, Klock, & Cook, 2006; Stanton, 2016).

Four reasons motivate us to pursue this work. First, we seek to articulate the sociotechnical concept of *stabilization* with respect to the more technocentric concept of *standardization* as an organizing principle for assembling the digital resources supporting distributed collaboration. We see stabilization as a negotiated order that brings together human and non-human participants into a more coherent arrangement. We contrast this to standardization or the reliance on technical standards to encourage interoperability among particular technological elements, with the people tethered (willingly or not) to such arrangements.

Second, as part of distinguishing stabilization from standardization, we articulate stabilization as being an example of distributed cognition, and seek to connect these literatures (e.g., Nemeth et al., 2006; Stanton, 2016). Scholars of human factors and particularly those focused on the complexities of distributed cognition involving material and digital arrangements have developed a rich and robust conceptual framework that bears directly on stabilization, distributed scientific work, and current thinking on cyberinfrastructure (CI) (Hutchins, 1995; Salas & Fiore, 2004; Fiore & Wiltshire, 2016; Nemeth et al., 2006; Stanton, 2016).

Our third reason for advancing stabilization practices as core to distributed work is that contemporary scientific endeavors are increasingly differing from traditional approaches to science in that projects are growing in terms of both budget and effort, resulting in larger and more distributed teams and a greater reliance on computing and other digital arrangements, typically referred to as CI (Atkins et al., 2003; Foster & Kesselman, 2003). There is now a visible body of testimonial literature about this trend (Atkins et al., 2003; Hey & Trefethen, 2003; Katz & Martin, 1997); a growing collection of empirical literature providing insights into movements toward CI-enabled science practice (cf. Ribes & Lee, 2010); a nascent area of literature providing conceptual insight into these practices (Jackson, Edwards, Bowker, & Knobel, 2007; Ribes & Finholt, 2009); and a history of various design approaches to designing and building the digital platforms supporting computer-enabled science (e.g., Borgman, Bowker, Finholt, Arbor, & Wallis, 2009; Jirotka, Lee, & Olson, 2013; Lee, Dourish, & Mark, 2006).

Our fourth reason for pursuing this work is that the burgeoning literature on CI seems to be predicated on an implicit model of science where roles are well-defined, scientific workflow is stable, discrete, and modular, and the value of these digital infrastructures lies in their provision of high-value assets (such as complex, rare, or sizable combinations of multiple data sets, analytic devices, and other massive-scale resources) to a large, distributed, and often diffuse number of scientists (Atkins et al., 2003). This model of scientific practice reflects a socially thin, compartmentalized, and streamlined version of scientific work. In contrast, research and reports on the observed collaboration practices of these scientists make clear that distributed work in this domain also depends on building long-term relationships of trust and mutual understanding (e.g., J. Cummings, Finholt, Foster, Kesselman, & Lawrence, 2008; Star & Ruhleder, 1996; Haythornthwaite, 2009). The empirically centered work presented here helps make clear that scientists using CI often rely on significant social supports and the sustained engagement of colleagues. That is, there is substantial social activity to move the scientific endeavor forward, what Haythornthwaite (2009) articulates as "heavyweight" social activity, requiring extensive negotiation and discussion, and reliant on building and sustaining shared understandings.

In this chapter we build from our studies of the work practices and uses of digital resources by social scientists collaborating at a distance in order to advance empirical and theoretical insights into the ways in which infrastructural stability—the efforts to ensure that digital resources and work practices stay aligned—support science practices. Specifically, we make the case that the sociotechnical concept of stabilization provides a more robust view of how scientists use CI than does the more well-known, and technology-centered, concept of standardization. Stabilization reflects the hard work of negotiation, adaptation, and shifting work practices that allow those using a particular technological system (in this case scientists using CI), to settle on a particular arrangement (see Russell & Williams, 2002). Stabilization embodies a balancing of competing technological and social forces—different possibilities of use, different goals of the users and designers, and emerging possibilities of use. In contrast, the concept of standardization focuses attention to the ways in which specific technological arrangements are designed to allow the elements to interoperate on a systems level. Standardization is what allows multiple devices to be connected together, and for documents to be visible and editable on different digital platforms. Standardization is a tremendous engineering feat, even as these often diminish the ways in which people organize and leverage the standards.

Building from this, we set out here to theorize on the role of stabilization in distributed work, and the ways in which these efforts are enmeshed into the digital and material technologies used by distributed teams. In the next section we lay out the literature relevant to studying social scientists' shared uses of CI. Then we lay out our research design, data collection, and data analysis. In the third section of this chapter, we highlight the forms of CI we observe scientists using. In the final section we explore implications of stabilizing for both research on scientific collaboration and for CI.

STUDYING USES OF CI BY SOCIAL SCIENTISTS

Social science enlists those whose central phenomenon of interest is the human condition, many of whom are found in disciplines such as: sociology, economics, anthropology, political science, linguistics, management and organizational studies, and others (Goff et al., 2011). Motivated by a gap in current research on the uses of CI to support science, we sought to explore the practices of social scientists involved in distributed collaborations. This exploration of social scientists' uses of digital resources enables us to compare the work of social scientists with what has been found in studies of natural, physical, and computer sciences (cited previously and detailed in the following sections), while also identifying variations among social scientists based on training, research methodologies, and personal preferences. And, the standardization of infrastructure within these CI environments is greatly influenced by the research practices, experiences, and uses of digital tools and material resources, and the data structures associated with diverse teams of scientists collecting and working with large data sets over extended periods of time (Ribes & Lee, 2010; Baker, Ribes, Millerand, & Bowker, 2005)

To do this we review three distinct areas of research. We begin by focusing on CI, acknowledging the implicit orientation towards CI's for the natural and physical sciences. Next, we describe the challenges of CI for distributed teams from a work practice perspective. In the third section we describe the enacted view of CI that focuses attention to the ways in which scientists currently learn and understand CI. Finally, we highlight what a practice-based perspective on distributed collaboration can contribute.

CI AND SCIENTIFIC WORK

Contemporary CIs are typically motivated by a need for some combination of (1) computational power, (2) data storage and access, and (3) shared uses of important (and often expensive) data collection, data analysis, and visualization platforms. Because of these needs, CIs often provide high-speed networking technologies to support computation, data access, and tool access. These CIs often are premised on large teams (i.e., dozens, if not hundreds, of scientists) with clearly defined roles, conducting distributed, collaborative research on pressing scientific problems. Increasingly, CI-enabled science is becoming the norm in the natural, physical, and biological sciences. Standardization of infrastructure within these environments is greatly influenced by the scientific practices and standardized data structures associated with diverse teams of scientists collecting and working with large data sets over extended periods of time (Ribes & Lee, 2010).

Current CIs in astronomy, bioinformatics, geosciences, ecology, medicine, and environmental studies have at least three things in common.[1] First, CI systems are typically designed for streaming data or the colocation of data to study a large-scale phenomenon. Second, CI-enabled work is long term and wide reaching with multiple stakeholders (Steinhardt & Jackson, 2015). Third, the goal of a CI is to support not just digital resources like data but also physical resources along with computationally intense environments for collaborative research work (Ribes, 2014). Fourth, CIs are supported by human infrastructures that manage the organizational and computational work needed to sustain them over time (Lee et al., 2006; Bietz, Baumer, & Lee, 2010).

Empirical insights from studies of CI uses make clear that much of the value of CI in these contexts is centered on gaining access to data. Baker, Ribes et al. (2005) argue that achieving data interoperability is the way to bridge intellectual communities. Likewise, sharing data on a large scale is seen as the key way to bring together interdisciplinary science to solve problems; this is intuitive but not well documented or systematically investigated (Borgman et al., 2009). Moreover, it appears that scientists are shifting their practices to leverage these resources. For example CI-enabled collaborations are increasingly being designed for leveraging large data sets, large-scale computing resources, and high-performance visualization tools, using collaborative workflows and allowing several people to share and work on the same analyses while located at different places (Jirotka et al., 2013).

Work Practices in Distributed Scientific Work

Distributed collaboration is increasingly common in scientific work (Atkins et al., 2003; Hey & Trefethen, 2005; Karasti, Baker, & Millerand, 2010) and lies at the heart of CI efforts (Borgman et al., 2009). Distributed collaborative scientific work refers to "interaction(s) taking place within a social context among two or more scientists that facilitates the sharing of meaning and completion of tasks with respect to mutually shared, subordinate goals" (Söderholm et al., 2008).

These efforts rely on complex social relationships that are often tightly interwoven into the scientific and technological practices of collaborators (e.g., Lee et al., 2006; Haythornthwaite, 2009). Scientists may choose to collaborate because they do not have sufficient skills or resources to pursue the work independently, they see value in drawing in other perspectives and talent, they are incentivized through funding and other reward structures to partner, or for other reasons

(Katz & Martin, 1997). Collaboration, more broadly, is fundamental to science as it is the basis of peer review and the epistemic cultures on which they rely (e.g. Knorr-Cetina, 1999, 2009)

Why does distributed scientific collaboration require complex social relationships? Beyond the technical difficulties of achieving data interoperability, Ribes and Lee (2010) identified four current challenges for developing CI to support distributed scientific collaboration: (1) heterogeneity of work practices and social structures among collaborators, (2) the need for standardization of data structures, work practices, and analytic approaches, (3) the importance of designing to allow changes over time, and (4) the need to design CI to be sustainable (and maintainable) over time scales of decades, what Ribes and Finholt (2009) call the "long now" of technological infrastructure (see also Karasti et al., 2010). Each alone is quite challenging; that all four of these, and the efforts to ensure data interoperability (e.g. Baker & Yarmey, 2009), are happening together magnifies the complexities involved in maintaining these collaborative practices.

Supporting scientists who work at a distance requires ICTs that make it possible to interact routinely, to easily access data repositories, and to support opportunities for connections with other scholars that overcome time and space (Olson & Olson, 2000; Crowston, Specht, Hoover, Chudoba, & Watson-Manheim, 2015). The evidence makes clear that coordinating these activities requires heavyweight social effort and comes at a substantial cost in both time and money (J. N. Cummings & Kiesler, 2008). This social effort is nonlinear, as a function of the number of dyads among all the project members, the level of turnover of members, and—to some extent—the complexity of the shared work. And, given both the changing nature of scientific enterprises and increasing levels of collaborative scientific effort, CI scholars emphasize the concept of the "long now" or the importance of designing for an open future (Ribes & Finholt, 2009). Changes over time in project institutional affiliation, membership, and work practices need to be accommodated in order for a CI to be sustainable across scientific inquiry (Bietz, Ferro, & Lee, 2012).

Jackson et al. (Jackson, Ribes, Buyuktur, & Bowker, 2011), building on a multi-year, multi-site, ethnographic field study of CI uses, found several issues related to time were also problematic in distributed collaborative scientific practice (see also Watson-Manheim, Chudoba, & Crowston, 2012). Jackson et al. observed that scientists had difficulty working with collaborators from institutions who have different academic calendars. More broadly, temporal rhythms of one's local work practices, organizational calendars, and scientific deadlines (for grants, conferences, and meetings) were critical shapers of distributed collaborative practices (e.g., Orlikowski & Yates, 2002).

A CI's sustainability rests on an ability to maintain standards, therefore one solution for addressing these challenges is standardization of data, work practices, and digital infrastructures (Ribes & Lee, 2010). Studies have highlighted the importance of data standardization and interoperability (Edwards, Mayernik, Batcheller, Bowker, & Borgman, 2011; Hey & Trefethen, 2005), while others have articulated the importance of, and difficulties with, standardizing scientific and analytic practices (e.g., Hara, Solomon, Kim, & Sonnenwald, 2003; Ribes & Lee, 2010). Still, the importance of supporting heavyweight social activities in distributed collaborations and addressing challenges related to standardization of CI in this domain is driving scholars to more carefully examine the evolving interdependencies among scientists, work practices, digital and material resources, and infrastructure.

CI Practices

We conceive of scientific work as emerging from enacted practices (i.e., the ways in which people *do* science in specific contexts) that are embedded within organizational and material structures including digital and material resources (see also Star & Ruhleder, 1994). Seen this way, infrastructure like CIs *are* practices: bound up in doing, not simply being present. That is, a CI is fundamentally a sociotechnical arrangement: a combination of material and digital resources entwined with the activity of doing science (Ribes & Lee, 2010). This practice-based perspective focuses attention on the *doing of infrastructure* in science work. As such, digital infrastructures are organic entities that emerge out of "information and work needs" (Feldman & Orlikowski, 2011; Østerlund & Carlile, 2005; Star, 2010).

Seen this way, CI are enacted, their digital and material properties become involved in science as the collective product of shared forms of practices and help to demarcate the specific ways scientists engage with one another. In related work, Sawyer, Crowston, and Wigand (2014) explore the computerization of real estate agents' work practices, focusing specifically on the ways real estate agents drew on various elements of existing digital infrastructures in ways that made sense to them. Sawyer et al. characterize the ways in which realtors perform digitally enabled work by

> drawing on many different computing elements and software based systems selected by individuals that may not be well-integrated or formally planned . . . [and] is thus ad hoc and only governed to the extent that the individual doing the assembling is making choices.
>
> (p. 51)

The resulting "infrastructure in use," while not reaching the level of formalization provided by a CI, is functionally equivalent to a standardized information system, but retains flexibility (Ribes & Polk, 2014) to respond to a community member's evolving work practices and material resources (Pollock & Williams, 2010; see also McNeese, Rentsch, Burnett, Pape, & Menard, 1998). As shown by the scientists in Vertesi's (2014) study and Sawyer et al.'s (2014) real estate agents the doing of infrastructure relies on learning practices from peers. This has less to do with learning formalized protocols or procedures and more to do with discovering ways to work *with* and *through* stabilized digital assemblages.

RESEARCH DESIGN, DATA COLLECTION, AND ANALYSIS

Because of our interests in advancing current empirical and conceptual understanding of social scientists' work practices and uses of CI resources, we designed an exploratory study (Punch, 2013). Our goal was theory elaboration: to advance the depth and clarity of current conceptualizations of cyberinfrastructure (e.g., Dutton, 2007; Ribes & Lee, 2010). Studies of scientific CI in use highlight that these digital resources are important and that standardization of data structures and research practices is one of the reasons for this. To advance this theorizing we designed the data collection in ways that engaged current scientists, that focused on work practices and the use of digital resources, and that focused on typically sized collaborative efforts (see also Sawyer, Østerlund, & Kaziunas, 2012).

Data Collection

Data collection methods consisted of a combination of semi-structured interviews and a detailed survey focused on digital tools and practices. The four-part semi-structured interview protocol was designed to be conducted face-to-face or via telephone. We asked respondents to (1) describe current distributed collaborations; (2) provide specific details about one of those projects; (3) describe current work practices on that project; and (4) detail the particulars of the digital elements (hardware, software, and services) used to support distributed collaborations. For each section, there were three to five broad questions with a number of probes.

We developed a survey to supplement the interview by capturing details of specific tools and patterns of use that might have otherwise been lost in narratives recounting specific projects, collaborative relationships, and challenges faced in conducting distributed social science work. The survey included a comprehensive "check sheet"—a form in which we asked participants to share information about collaborators as well as indicate in some detail specific digital devices, software tools, and other infrastructural elements that they used. The interview protocol and survey were designed so that the interviewee could work on the check sheet while we worked through the interview questions. The survey was available both on paper and online, with delivery mode dependent on the resources available at the time of interview and preferences of respondents.

Relying on several rounds of snowball sampling, we conducted interviews with 31 social scientists across one calendar year. Interviews took 70 minutes, on average. Interviewing took place in two extended phases (of 14 and then 8 interviews) separated by several months of interim analysis. During the interview, each social scientist shared details about a single distributed collaboration in which he or she was involved. Reports of collaborative practices provided to us by our primary informants gave us insights into the work practices of over 170 researchers and project staff (with all but two working on teams of between four and seven people). The 31 interviews represent the experiences of 8 men and 14 women pursuing social science questions in the intellectual communities of science and technology studies, information science, information systems, and HCI/CSCW. Nine of the 31 were tenure-track faculty and 14 were tenured faculty at research-intensive institutions in the United States and abroad; eight were research faculty or post-doctoral researchers. We note this distribution of experience (and likely experience) in our discussion of the findings in the next section.

Data Analysis

We recorded and transcribed the interviews, made digital copies of the check sheet to support analyses, and gathered particular documents and other project material (such as websites) when allowed. Transcripts of the 31 interviews were analyzed inductively to identify common themes related to particular aspects of (1) work practices of collaborative teams and (2) the ways in which these teams interacted with their CI and other digital technologies. To pursue the first focus, we began with a nominal list of practices and modified the initial list based on the data. To pursue the second focus, we coded the details of particular technologies being used as a means of identifying patterns of technological arrangements that were common to multiple teams. We also focused on the ways in which participants articulated uses of CI and technologies as they discussed practices and issues.

Two members of the research team coded each interviews and then worked through several iterations of a standardized coding scheme with the entire research team to clarify emergent themes, resolve differences, better articulate formative codes, and add new concepts to the analytic scheme. Typical of open coding on interviews and field notes, during this data analysis, consensus was valued more highly than inter-rater reliability (Glaser & Strauss, 1967).

TWO FINDINGS: NASCENT CI AND STABILIZING PRACTICES

We focus on stabilization practices, a topic that emerged from the data (it was not one of the initial practices being coded). Stabilization took the form of efforts to adapt to changes in the features and functions of specific software, issues with compatibility across different software applications and suites, changes in research collaboration member's collective or individual expertise with software (due to changes in personnel or changes in software), shifts in team member's understanding of the working arrangements, learning effects when decisions were made to switch software (for a variety of reasons), and how people recovered from mistakes with version control and document management.

To better understand these stabilization efforts, we next report on patterns of digital technology uses to make clear four insights from the data. First, data show there is no standard arrangement, no singular set of digital technologies, being used by respondents. Second, data make clear that there are many combinations and overlapping uses of similar software and applications, that these arrangements evolve for a number of reasons, and that this change and overlap is expected by the participants. Third, data make clear that interoperability—enabled by standards—among the software, applications, and platforms relies on many technical standards (such as CSV and XML for sharing data, PDF and other standard document formats for sharing files, and many standards for email, web, and internet working). Fourth, we observe that these collections of digital technologies, while having similar functionality, being comprised of multiple commercial and open-source applications and platforms, and all being used to support distributed scientific collaboration, fall short of being an enduring CI. As such, standardization is a necessary technological requirement for these proto-CI arrangements, but is not sufficient to ensure these digital arrangements are workable or useful for the collaborating scientists.

As readers might expect, participants reported having and using several computers (both laptops and desktops), tablets, and smartphones in support of their work. They used these various devices at home, at various places at work, in public spaces, and in between. Several had print servers at home. Most digital devices were provided by an employer, although each respondent noted they also owned several personal devices. And, respondents indicated that they routinely used both personal and employer-provided devices together. Perhaps unsurprisingly, email was the dominant means for interacting, collaborating, sharing materials, and for almost all project work. While many of our respondents were active in social media, they reported that email was still the common denominator, omnipresent in their lives and accessed across multiple devices: "A lot of people still email things around even though there's a Dropbox" (JMp6).

These digital resources were used every day, often multiple times. In describing their uses of devices, most respondents found it challenging to separate activities directly related to the specific collaborative project being described, similar activities being performed for different projects, and personal use. This noted, only some of these uses pertain directly to their distributed project collaborations. However, emails pertaining to student grading, service assignments, and data analysis arrive to the same device. What this means is that scientific practices and access to digital scientific resources were embedded into the larger sphere of their daily life and both science and non-science happened across the same set of digital and material resources, mixed together across time. None of the 31 respondents reported having distinct labs as the site of their research. While most had one or two preferred places to work (their office at home or a favorite cafe), in practice their research happens at different times in their personal schedules and across many physical places, even in public spaces like airports as they travel. Collectively, through doing these practices, scientists enact and recreate the shared distributed cognition that defines how they work.

Respondents identified using nearly 70 distinct software applications and platforms. These included common commercial software and hardware such as Microsoft (MS) Office applications, the Google suite of software, standard browsers such as Firefox or Safari (but rarely Internet Explorer), typically running on Microsoft or Apple computers. We note that all of these are either commercial products or open-sourced; none are purpose-built CI components. The teams and collaborators who provided us data relied on commercial and community offerings, a "commodified" approach to technology development that relies on standards and, for the commercial products, market competition (Sawyer, 2000).

Most of these commercial products provide a means to share data and documents across rival platforms. This means that team members can choose to use different word processing software and exchange files, or to experiment with particular cloud-storage providers, or move from SPSS to *R*, knowing that these are more or less compatible, more commodities than specialized.

Nineteen respondents report using Gmail; 21 also use a second email address and/or client (such as Thunderbird, Outlook, or Apple Mail). Most are using no more than two email clients, though there are 67 different email uses across the 31 respondents. All but one respondent use MS Word, and 22 use Google Docs, which means that most respondents use both. Twenty-seven respondents reported using some form of reference management software (e.g., Endnote (12), Zotero (8), Mendeley, and others), with several people using more than one. This noted, no one reference management software application had a majority of use. Likewise, there was a wide range of analytic software being used (with R having nine counts; SPSS, SAS, Atlas TI, and Dedoose all having four counts; and many with one count). If a respondent reported using analysis software, they reported more than one. Only three of 31 interviewees mentioned doing collaborative data analysis (on numerical data); the rest participated in collaborations where each individual did their own analysis and shared or pooled their results.

All respondents reported using one or more cloud-based repository such as Dropbox, Google Drive, or Box to store and share project documents. No respondent relied on his or her institution's learning management system or on private infrastructures (like personal web space) to serve this function.

Skype served primarily as a vehicle for project meetings during which updates, to-dos, and plans were discussed, though ten people also noted they used call-in conference phone services. Participants reported using a range of project management software (but only by 13 of the respondents). Most had one use. Some reported using note-taking software (e.g., Evernote). Five participants reported using GitHub (in the project management category).

Participants spoke at length of their efforts to make the various elements of the proto-CI in use work. Some of the issues reflect the realities of having to know many pieces of software. For example, having to discuss how to move numerical data from Excel to R, or how best to format a Google-Doc-based document so that it can be copied into a Word document. This form of articulation work is common outside of scientific collaboration and independent of a reliance on CI. Stabilization differs from articulation in that it requires a collective negotiation: multiple participants are involved and the outcome or decision requires the team to adhere to the new arrangements (whereas articulation work is simply extra work, often done by one person outside the view of others).

Stabilization practices show up in both the mundane activities of working together, often most visible when personnel change and issues arise. Two constants of these collaborative efforts, reported on by all 31 respondents, are (1) the steady flow of students (both graduate and undergraduate) on and off of the research teams and (2) collective attention to reflecting on how the collaborators are following protocols and using specific digital technologies. The steady changes in team composition lead to constant discussions about basic procedures (how do we name files, where are they placed, who takes notes, who manages the to-do list) and larger tasks and goals and a re-establishment and redistribution of shared cognition (Hutchins, 1995; Salas & Fiore, 2004; Fiore & Wiltshire, 2016; Stanton, 2016). Relative to the basic procedures, the proto-CI elements are involved in subtle and not-so-subtle ways: updating which citation management system to use based on losing a high-skilled student for one who has limited skills on one application; getting students to use Dropbox instead of Google Docs; or convincing new graduate students to share notes in a common folder.

Relative to stabilization efforts around larger tasks and goals, the entire team gets involved when a senior doctoral student moves on, as they collectively must re-sort who does what and which digital platforms to use. Stabilization efforts often frame decisions on projects, tasks, and to-dos; it is an omnipresent effort.

STABILIZING NOT STANDARDIZING: IMPLICATIONS FOR RESEARCH AND PRACTICE

Through this research we come to recognize that customized assemblages of standardized tools create a *stabilized* (but not static) infrastructure that enables collaborative teams to maintain flexibility in their work practices throughout the evolution of their research project. Second, this shift of focus from standardization to stabilization in the context of the heavyweight collaborative activities that lead to distributed cognition has implications for how we think about infrastructure design to support scientific work. Designs for CI need to enable scientists to build and maintain shared mental models (e.g., Stanton, 2016).

These observations echo findings from other empirical studies of CI use and distributed scientific collaboration (Fry, 2006; Lin, Procter, Halfpenny, Voss, & Baird, 2007; e.g., Ribes & Lee, 2010). The social scientists in our study described their efforts to *assemble*, rather than standardize, their digital infrastructures to accommodate their heterogeneous—and often idiosyncratic—preferred work and social practices. Teams relied on commodified software and commercial digital networks and storage providers, working digital infrastructures, relatively short time horizons, and stronger commitments to relationships and practices over specific hardware or software. These collections of commodity technologies point to stable, but ad hoc, sets of arrangements, more reflective of the digital assemblages of real estate agents discussed previously (Sawyer et al., 2014) than of CI. These bundles of ICT are not an information system in the classic sense: there is no governance, no defined coherence; rather, functional equivalence driven more by the similarity across these respondents' work roles and practices than by any a priori design (Sawyer et al., 2014).

Given these observations, we argue that the practice of managing the confluence of infrastructures and collaborative practices that we see in the work of distributed social scientists is more closely associated with technological *stabilization* (Pinch & Bijker, 1987) than *standardization*. In discussing mechanisms of technology development, Pinch and Bijker challenge the idea that technologies develop as a result of a predetermined march towards standardization. In Pinch and Bijker's (1987) conceptualization of stabilization, new technologies undergo a period of interpretive flexibility marked by multiple options and incarnations followed by a phase of social integration that ultimately results in a stabilizing convergence around a particular arrangement. They provide the example of the development of the bicycle. Some early bicycle designs failed and others thrived. Success or failure was not based on a linear evolutionary model where one design arose from the previous, but rather as a result of multidirectional experiments that enabled makers to explore a range of different options within the two-wheeled peddled vehicle design space, each responding to distinct problems faced by different user groups. Eventually designs stabilized into the safety bicycle (two wheels of equal size, air tires and a seat in a more or less forward position), a form that has remained relatively constant ever since.

We recognize in this description the assemblages of the social scientists in our study; different components of a digital infrastructure are gathered in response to specific problems being faced by a team at a given time in their research and collaborative process. The beginning phases of a project reflect the interpretive flexibility described by Pinch and Bijker (1987) during which different tools and practices are evaluated and considered. This is followed by a stabilizing convergence during which a team settles on a given configuration of tools. However, where Pinch and Bijker (1987) talk about the process of technological stabilization as being completed once and done, we saw evidence of the teams in our study returning to this process repeatedly during the course of a project's lifetime, re-interpreting or re-negotiating the specifics of their infrastructure assemblage in response to changes in research practices, personnel, and/or resources.

In this sense, the stabilizing practices of the social scientists in our study evoke Vertesi's (2014) "seamful spaces." She draws a compelling contrast between the goal of "seamless integration" envisioned by ubiquitous computing researchers and the reality of "seamful" digital environments in which many distributed scientific collaborations exist. In practice, social scientists are continually assessing and responding to changing infrastructural needs at technoscientific, sociotechnical, and institutional levels, three dimensions of change described by Ribes and Polk in their discussion about infrastructure flexibility (2014). We examine this process in more detail next.

Ribes and Lee describe the ways in which CI supports science "by identifying vast interdisciplinary swaths that could benefit from data and resource sharing, knowledge transfers, and support for collaboration across geographical, but also institutional and organizational divides" (Ribes & Lee, 2010, p. 232). We previously highlighted some of the challenges that CI designers face when trying to accomplish this, including: (1) the heterogeneity of work practices and social structures among collaborators, (2) the need for standardization of data structures, work practices, and analytic approaches, (3) the importance of designing to allow changes over time, and (4) the need to design CI to be sustainable (and maintainable) over long time periods (e.g., multiple decades).

Based on what is reported in current CI research, we anticipated that our respondents would use standardized tools in consistent ways across projects because they provided the necessary digital infrastructure to support their social science practices. However, as we learned more about the work practices of these distributed collaborating scientists, we discovered that standardized tools and infrastructures, such as email, Google Docs, and Dropbox, were often used in different ways by each team of researchers. The science being conducted by our participants' teams required sustained engagement, flexibility, and responsiveness to fellow team members. We identified this as a culture of accommodation, similar to what Schroeder and Spencer (2009) have previously reported: social scientists are both inventive and adaptable. We observed a high degree of interdependence among the processes for defining roles, negotiating rules or routines, and articulating tasks. Many of the stabilizing and adaptive actions described were motivated by a need to support valued heavyweight aspects of teamwork better conceptualized as distributed cognition

(Hutchins, 1995; Salas & Fiore, 2004; Stanton, 2016). Overall, the respondents' individual science practices were loosely and often uniquely mapped to the shared practices of a specific distributed scientific collaboration (see also McNeese et al., 1998).

We also note that others have found that sustaining data access and interoperability is difficult for CI (e.g., Baker, Jackson, & Wanetick, 2005; Bietz et al., 2012; Lawrence, 2006; Lee et al., 2006; Barjak et al., 2009); therefore, we highlight the opportunity to develop tools that will help transition some research tasks to lighter-weight environments through responsive workflow support (e.g., Fry, 2006). Further, training regarding how to collaborate is typically learned in apprenticeship, by watching faculty move from project to project and mimicking this craft as part of scholarly maturation. In order to create more intentionally around these adaptive and stabilizing practices, it will become increasingly important to train PhD students on the fundamentals of distributed collaboration, to help them learn community or shared data practices (e.g., 2007) and to develop basic technical skills that will enable them to practice adaptive design to modify and optimize their collaborative environments. Part of this training will need to involve cultivating the ability to assess, implement, and evaluate the effectiveness of digital tools in terms of functional equivalence.

ACKNOWLEDGMENTS

Funding provided by National Science Foundation (NSF) via Grant CI 1527410. The findings and opinions reported are solely the authors and do not represent or speak for the NSF.

We gratefully acknowledge the time and insights provided by so many scholars, and the comments from reviewers and audience members of the various talks, panels, and feedback sessions we have held. Thank you to Michael D. McNeese and Eduardo Salas for specific comments on earlier versions of this chapter.

We thank the many students who have contributed so much to this program of study over the years, to include: post-doctoral fellow JoAnn Brooks; doctoral student Sarah Bratt; master's students Elizabeth Kaziunas, Doratea Szkolar, Michelle Brown, Megan Threats, Elizabeth Greenberg, and Nirihika Ved; undergraduate student Nicholas Kapteyn; and post-baccalaureate research fellow Anjelica Torcivia.

NOTE

1. This is an illustrative list; there are many forms of CI to be found in many intellectual spaces not mentioned.

REFERENCES

Atkins, D. E., Droegemeier, K. K., Feldman, S. I., Garcia-molina, H., Klein, M. L., Messerschmitt, D. G., . . . Wright, M. H. (2003). Revolutionizing science and engineering through cyberinfrastructure: Report of the national science foundation blue-ribbon advisory panel on cyberinfrastructure. *Science*, *81*(8), 1562–1567. Retrieved from www.nsf.gov/od/oci/reports/atkins.pdf

Baker, K. S., Jackson, S. J., & Wanetick, J. R. (2005). Strategies supporting heterogeneous data and interdisciplinary collaboration: Towards an ocean informatics environment. In *HICSS'05*. New York: IEEE Computer Society Press.

Baker, K. S., Ribes, D., Millerand, F., & Bowker, G. C. (2005). Interoperability strategies for scientific cyberinfrastructure: Research and practice. *Proceedings of ASIST, 42*(1).

Baker, K. S., & Yarmey, L. (2009). Data stewardship: Environmental data curation and a web of repositories. *International Journal of Digital Curation, 4*(2), 12–27.

Barjak, F., Lane, J., Kertcher, Z., Poschen, M., Procter, R., & Robinson, S. (2009). Case studies of e-infrastructure adoption. *Social Science Computer Review, 27*(4), 583–600.

Bietz, M. J., Baumer, E. P., & Lee, C. P. (2010). Synergizing in cyberinfrastructure development. *Computer Supported Cooperative Work (CSCW), 19*(3–4), 245–242.

Bietz, M. J., Ferro, T., & Lee, C. P. (2012). Sustaining the development of cyberinfrastructure: An organization adapting to change. *CSCW'12*, 901–910.

Borgman, C. L., Bowker, G. C., Finholt, T. A., Arbor, A., & Wallis, J. C. (2009). Towards a virtual organization for data cyberinfrastructure. In *Proceedings of the 9th ACM/IEEE-CS joint conference on digital libraries* (pp. 353–356). New York: IEEE Computer Society Press.

Crowston, K., Specht, A., Hoover, C., Chudoba, K. M., & Watson-Manheim, M. B. (2015). Perceived discontinuities and continuities in transdisciplinary scientific working groups. *Science of the Total Environment, 534*, 159–172 . https://doi.org/10.1016/j.scitotenv.2015.04.121

Cummings, J. N., Finholt, T., Foster, I., Kesselman, C., & Lawrence, K. A. (2008). *Beyond being there: A blueprint for advancing the design, development, and evaluation of virtual organizations*. Report from an NSF Workshop on Developing Virtual Organizations.

Cummings, J. N., & Kiesler, S. (2008). Who collaborates successfully ? Prior experience reduces collaboration barriers in distributed interdisciplinary research. *CSCW'08*, 437–446. https://doi.org/10.1145/1460563.1460633

Dutton, W. H. (2007). Reconfiguring access to information and expertise in the social sciences: The social shaping and implications of cyberinfrastructure. In *Proceedings 3rd international conference on e-social science, October 7–9*. Ann Arbor, MI: The University of Michigan. Retrieved from http://ess.si.umich.edu/papers/paper129.pdf

Edwards, P. N., Mayernik, M. S., Batcheller, A. L., Bowker, G. C., & Borgman, C. L. (2011). Science friction: Data, metadata, and collaboration. *Social Studies of Science, 41*(5), 667–690.

Feldman, S. I., & Orlikowski, W. (2011). Theorizing practice and practicing theory. *Organization Science, 22*(5), 1240–1253.

Fiore, S., & Wiltshire, T. (2016). Technology as teammate: Examining the role of external cognition in support of team cognitive processes. *Frontiers in Psychology, 7*, 15–31. https://doi.org/10.3389/fpsyg.2016.01531

Foster, I., & Kesselman, C. (2003). *The grid 2: Blueprint for a new computing infrastructure*. Amsterdam, Netherlands: Elsevier.

Fry, J. (2006). *Coordination and control of research practices across scientific fields: Implications for a differentiated e-science*. Hershey, PA: Ed. Info. Sci. Pub.

Glaser, B. G., & Strauss, A. L. (1967). *The discovery of grounded theory: Strategies for qualitative research*. Hawthorne, NY: Aldine de Gruyter.

Goff, S. A., et al. (2011). The iPlant collaborative: Cyberinfrastructure for plant biology. *Frontiers in Plant Science, 2*(34).

Hara, N., Solomon, P., Kim, S., & Sonnenwald, D. H. (2003). An emerging view of scientific collaboration: Scientists? perspectives on collaboration and factors that impact collaboration. *JASIST, 54*(10), 952–965.

Haythornthwaite, C. (2009). Crowds and communities: Light and heavyweight models of peer production. In *HICSS'09* (pp. 1–10). New York: IEEE Computer Society Press.

Hey, T., & Trefethen, A. (2003). e-Science and its implications. *Philosophical Transactions of the Royal Society of London Series A Mathematical Physical and Engineering Sciences, 361*(1809), 1809–1825. Retrieved from http://eprints.ecs.soton.ac.uk/7964/

Hey, T., & Trefethen, A. (2005). Cyberinfrastructure for e-Science. *Science, 308*(5723), 817–821. Retrieved from http://eprints.soton.ac.uk/260844/

Hutchins, E. (1995). *Cognition in the wild.* Cambridge, MA: MIT Press.

Jackson, S. J., Edwards, P. N., Bowker, G. C., & Knobel, C. P. (2007). Understanding infrastructure: History, heuristics, and cyberinfrastructure policy. *First Monday, 12*(6).

Jackson, S. J., Ribes, D., Buyuktur, A., & Bowker, G. C. (2011). Collaborative rhythm: Temporal dissonance and alignment in collaborative scientific work. In *CSCW'11* (pp. 245–254). New York: Association of Computing Machinery (ACM) Press.

Jirotka, M., Lee, C. P., & Olson, G. M. (2013). Supporting scientific collaboration: Methods, tools and concepts. *CSCW'13*, 1–49. https://doi.org/10.1007/s10606-012-9184-0

Karasti, H., Baker, K. S., & Millerand, F. (2010). Infrastructure time: Long-term matters in collaborative development. *Computer Supported Cooperative Work (CSCW), 19*(3/4), 377–415.

Katz, J. S., & Martin, B. R. (1997). What is research collaboration? *Research Policy, 26*(1), 1–18.

Knorr-Cetina, K. (1999). *Epistemic cultures.* Cambridge, MA: Harvard University Press.

Knorr-Cetina, K. (2009). *Epistemic cultures: How the sciences make knowledge.* Cambridge, MA: Harvard University Press.

Lawrence, K. A. (2006). Walking the tightrope: The balancing acts of a large e-research project. *Journal of Computer Supported Cooperative Work, 15*(4), 385–411.

Lee, C. P., Dourish, P., & Mark, G. (2006). The human infrastructure of cyberinfrastructure. *CSCW'06*, 483–492. https://doi.org/10.1145/1180875.1180950

Lin, Y.-W., Procter, R., Halfpenny, P., Voss, A., & Baird, K. (2007). An action-oriented ethnography of interdisciplinary social scientific work. In *Proceedings of the 3rd e-Social science conference, October 7–9.* Ann Arbor, MI: The University of Michigan. Retrieved from http://ess.si.umich.edu/papers/paper179.pdf

McNeese, M., Rentsch, J., Burnett, D., Pape, L., & Menard, D. (1998). *Testing the effects of team processes on team member schema similarity and team performance: Examination of the team member schema similarity model.* United States Air Force Cognitive Systems Division Report AFRL-HE-WP-TR-1998-0070, 45pp.

Nemeth, C., O'Connor, M., Klock, P. A., & Cook, R. (2006). Discovering healthcare cognition: The use of cognitive artifacts to reveal cognitive work. *Organizational Studies, 2*, 1011–1035. https://doi.org/10.1177/0170840606065708

Olson, G. M., & Olson, J. S. (2000). Distance matters. *Human-Computer Interaction, 15*(2), 139–178.

Orlikowski, W. J., & Yates, J. (2002). It's about time: Temporal structuring in organizations. *Organization Science, 13*(6), 684–700.

Østerlund, C., & Carlile, P. (2005). Relations in practice. *The Information Society, 21*(2), 91–107.

Pinch, T. J., & Bijker, W. E. (1987). The social construction of facts and artifacts: Or how the sociology of science and the sociology of technology might benefit each other. In W. E. Bijker, T. P. Hughes, & T. Pinch (Eds.), *The social construction of technological systems: New directions in the sociology and history of technology* (pp. 17–50). Cambridge, MA: MIT Press.

Pollock, N., & Williams, R. (2010). E-infrastructures: How do we know and understand them? Strategic ethnography and the biography of artefacts. *Computer Supported Cooperative Work (CSCW), 19*(6), 521–556.

Punch, K. F. (2013). *Introduction to social research: Quantitative and qualitative approaches*. Thousand Oaks, CA: Sage.

Ribes, D. (2014). Ethnography of scaling, or, how to a fit a national research infrastructure in the room. In *Proc CSCW* (pp. 158–170). New York: Association of Computing Machinery (ACM) Press.

Ribes, D., & Finholt, T. A. (2009). The long now of technology infrastructure: Articulating tensions in development. *Journal of the Association for Information Systems*, *10*(5), 5.

Ribes, D., & Lee, C. P. (2010). Sociotechnical studies of cyberinfrastructure and e-Research: Current themes and future trajectories. *JCSCW*, *19*(3–4), 231–244. https://doi.org/10.1007/s10606-010-9120-0

Ribes, D., & Polk, J. B. (2014). Flexibility relative to what? Change to research infrastructure. *Special Issue of Journal of the Association for Information Systems*, *15*, 287–305.

Russell, S., & Williams, R. (2002). Social shaping of technology: Frameworks, findings and implications for policy with glossary of social shaping concepts. In K. Sorensen & S. Russell (Eds.), *Shaping technology, guiding policy: Concepts, spaces and tools* (pp. 37–132). London: Edward Elgar.

Salas, E., & Fiore, S. M. (Eds.). (2004). *Team cognition: Understanding the factors that drive process and performance*. Washington, DC: American Psychological Association.

Sawyer, S. (2000). Packaged software: Implications of the differences from custom approaches to software development. *European Journal of Information Systems*, *9*(1), 47–58.

Sawyer, S., Crowston, K., & Wigand, R. T. (2014). Digital assemblages: Evidence and theorising from the computerisation of the US residential real estate industry. *New Technology, Work and Employment*, *29*(1), 40–56.

Sawyer, S., Østerlund, C., & Kaziunas, E. (2012). Social scientists and cyberinfrastructure: Insights from a document perspective. In *CSCW'12, February, 11–15*. New York: Association of Computing Machinery (ACM) Press.

Söderholm, H. M., Sonnenwald, D. H., Cairns, B., Manning, J. E., Welch, G., & Fuchs, H. (2008). Exploring the potential of video technologies for collaboration in emergency medical care: Part II. Task performance. *JASIST*, *59*(14), 2335–2349.

Stanton, N. (2016). Distributed situation awareness. *Theoretical Issues in Ergonomics Science*, *17*(1), 1–7.

Star, S. L. (2010). This is not a boundary object. *STHV*, *35*(5), 601–617.

Star, S. L., & Ruhleder, K. (1994). Steps towards an ecology of infrastructure: Complex problems in design and access for large-scale collaborative systems. *CSCW'94*, 253–264. https://doi.org/10.1145/192844.193021

Star, S. L., & Ruhleder, K. (1996). Steps toward an ecology of infrastructure: Design and access for large information spaces. *Information Systems Research*, *7*(1), 111–134.

Steinhardt, S. B., & Jackson, S. J. (2015). Anticipation work: Cultivating vision in collective practice. *Proc CSCW*, 443–453.

Vertesi, J. (2014). Seamful spaces: Heterogeneous infrastructures in interaction. *Science, Technology & Human Values*, *39*(2), 264–284.

Watson-Manheim, M. B., Chudoba, K. M., & Crowston, K. (2012). Perceived discontinuities and constructed continuities in virtual work. *Information Systems Journal*, *22*(1), 29–52.

3 Collaborative Board Games as Authentic Assessments of Professional Practices, Including Team Cognition and Other 21st-Century Soft Skills

Michael F. Young, Jonna M. Kulikowich, and Beomkyu Choi

CONTENTS

LEARNING FROM PLAY

That we all learn from play seems undeniable. But often it takes an educational psychologist to point out the obvious: that for most of life's important lessons we as individuals, and as teams, learn first, and perhaps primarily from playing around (Vygotsky, 1966, 1978). Playful learning both relies on distributed team cognition and contributes to its development. While playful learning is a large part of childhood, it also remains a lifelong aspect of formal and informal education. Children play house without the burdens of really cooking, cleaning, or paying bills, or play cowboys without the dangers of real horses, guns, and stampeding animals.

Many simulations involve playful learning and draw on role playing and imagination to establish realistic, if not real, conditions for learning. Distributed cognition is an essential component of successful team play, whether in formal team sports or in massively multiplayer online games such as World of Warcraft. In such cases, key roles are predefined (quarterback vs. tight end, tank vs. healer) but success is largely determined by a shared and coordinated understanding of goals and execution of well-synchronized creative actions.

Yet, important dimensions of playful learning may be lost when play becomes the focus of experimental controlled studies, academic theory, scholarly analysis, assessment, and other research. Traditional approaches to measurement and research methods push for the assessment of isolated and individual play behaviors (not teams), linear conceptions of knowledge, and dispositions (not distributed knowledge) over discrete short periods of play from only single game instances (Resnick & Resnick, 1992; Roth, 1998; Schwartz & Arena, 2013). This stands in contrast to our real-life experiences where we play the same games repeatedly with different opponents and even with different strategies and different team roles, and learn not from single instances of game play, but from repeated experiences over substantial time, and through post-game discussion, reflection, and analysis with other players—referred to as metagame experiences. The dynamics of game play over various occasions and metagame reflections and interactions are an important aspect of playful learning that are often missed but are an essential element of our ecological and situated approach (Young, 2004; Young et al., 2012). Play is both something we do when we follow the rules of a game and an approach we take when entering into an interaction perhaps playfully testing the limits of those rules. This latter idea, of a playful approach to the world, represents an individual or group's goal set to intentionally manipulate aspects of the situation in order to achieve a playful interaction (Young & Slota, 2017).

As children play games, they learn physical and mental skills (Resnick, 2006) and acquire new way to affect the world (effectivities). They also learn strategies for how to play games and how to interact with other players (Gee, 2003; Squire, 2008). As described by a quote from a future teacher in our college teacher preparation program,

> When I was a child, I played all kinds of board games and physical games (like twister or spud). Those games taught me a lot of rules. For instance, I had to learn the rules for the game to play, but I also learned ethical rules, such as don't cheat and don't touch other people's cards/property. Physical games and games we played outside taught me basic coordination and how to play on a team. Some games were centered around educational concepts like counting (Hershey's Kisses game), or making words, or describing a word (Scrabble, Scattergories, and Taboo). Some required deductive reasoning and critical thinking (Clue), while others required drawing (Pictionary) or acting (charades).
>
> **Rebekah Labak, threaded discussion post, fall 2018**

It seems from games children can learn many of the basic social interactions they will need the rest of their lives. From games like Scrabble they can learn to be playful with language and add to their vocabularies with a focus on word relationships and spelling. In many games, like Monopoly, they can learn about money, value, and counting. And they can learn the basics of how to treat each other, including how to follow rules or how to cheat (hopefully not), as well. In this process they can playfully explore a personal ethic (strategy) as well as experience the social consequences of defying the rules. Play has also been linked to the development of creativity and other key aspects of child development (Howard-Jones, Taylor, & Sutton, 2002; Lillard, 2013; Resnick, 2008; Russ, 2003). It is interesting to note that for some approaches to early childhood education, like that of Maria Montessori, the fantasy aspects of play were decoupled from the mechanics of games and playful learning, and this remains an important distinction often discussed when trying to parse simulations versus games.

For our purposes, the question then becomes, what can adults learn from game play, and can game play be as powerful a social learning environment for advanced skills, such as the distributed team cognition of surgical teams debriefing or military planners during war games, as it is for development of early childhood social skills?

One important distinction to make at the outset of this chapter is between structured play, as is the case with most video games, card games, and board games, versus unstructured or "make believe" play. The latter form of playful learning has been discussed by child psychologists with regard to the development of social skills, creativity, and self-regulation (see e.g., Bodrova, Germeroth, & Leong, 2013; Lillard et al., 2013). In this chapter, we will focus on structured play that fits the standard definition of "game," which generally includes the presence of rules, turn-taking, and some sort of end goal state such as winners and losers.

Games have certainly been suggested as a teaching tool for skills in many advanced disciplines beyond preschool and early elementary school, and appear to be useful for learning about real-world environments including transportation planning (Huang & Levinson, 2012), safety science (Crichton, 2009), 5th grade science (Wilkerson, Shareff, Laina, & Grave, 2018), engineering design (Hirsch & McKenna, 2008), cybersecurity for avoiding phishing (Arachchilage & Love, 2013), and notably for our present topic, engineering teamwork (Hadley, 2014). Meta analyses of the effectiveness of games for learning advanced school content generally find small but positive effects of playful learning approaches (Clark, Tanner-Smith, & Killingsworth, 2015; Vogel et al., 2006; Wouters, van Nimwegen, van Oostendorp, & van der Spek, 2013; Young et al., 2012), but such effects are summarized across a diverse set of games and players and thus are hard to interpret at such a broad level of analysis. We have characterized this as looking for our princess in the wrong castle (Young et al., 2012) and have made a case that a more situated analysis of game play is required to account for the rich social context in which game play emerges.

In our work to study games as assessments, distinctions between what makes a game a simulation, or what makes a simulation also a game, have not been useful. At times, simulations are judged by their fidelity and verisimilitude, whereas games are judged by their playfulness. Yet upon closer inspection, games and more specifically game mechanics have been characterized variously as a subtype of simulations, as the superset of simulations, and parallel with shared element with simulations. For example, when developing taxonomies of games, Wilson et al. (2009) described

18 categories of game features, which Bedwell, Pavlas, Heyne, Lazzara, and Salas (2012) refined into nine features. Koehler, Arnold, Greenhalgh, and Boltz (2017) tested the value of these nine features for characterizing how gamers review and evaluate games, only to find that some additional features were needed. Drawing on Malone and Lepper's (1987) description of what makes learning "fun," many of these game features describe things that players enjoy (e.g., competition, conflict, control, human interaction) and not features that might distinguish key differences between the more veridical nature of simulations (real-world physics constraints) and the sometimes more fantasy-based nature of games. In our own work, we have found it best to characterize both simulations and games along the dimensions of playfulness, emotional experience as fun, and rules/game structures. For example, simulations (constrained by real-world parameters) can be playful, in the sense they allow users to explore openly and combine actions in ways not fully anticipated by instructors/designers. Likewise, many games involve some real-world constraints (gravity, solid structures) and we are aware of elite soccer players who use soccer video games to simulate player strategies. In the case of professional sports, playing a game may not be experienced as fun at all and be instead hard work. While some believe all games are a subset of simulations because they use some real-world constraints, others have argued that simulations are a subset of games because they use various game mechanics to various degrees and are sometimes experienced as fun and playful, like some games. For our purposes, none of these attempts to apply a taxonomy to games or organize the features of games in contrast to simulations have proved useful for research. Instead, we would assert that both games and simulations utilize various degrees of playful mechanics, are experienced with varying degrees of "fun," and have various levels of rules and end states/goals, whether competitive or collaborative or some combination (such as games with a betrayal mechanic).

For the topic of team distributed cognition, our interest is particularly on collaborative board games in which players perform game tasks together through communication, coordination, and other team process behaviors and achieve results as an integrated unit. There is an element of competition still with such games as they often include individual player scores, but collaborative board games are played against the board itself and the win condition is win or lose for all. For example, in Forbidden Island, time is running out as the island is sinking and players must collaborate and leverage their individual skills to escape before the water rises to cover the island; if the island sinks, the players all lose as a team.

As mentioned regarding engineering teamwork, Hadley (2014) used the board game Pandemic as an educational intervention to encourage engineers to reflect on their teamwork skills and strategies. In this study, board game play was the intervention, not the assessment. But this intervention highlights the need to focus work concerning team effectiveness on the interactions of player and board position dynamics rather than on a static momentary snapshot of individual scores, team scores, or individual cognitive strategies. For our purposes team cognition involves:

1. Intellectual diversity
2. Civil discourse, active listening

3. Balanced contributions across team members
4. Shared group decision making, consensus building
5. Coordinated goal/task planning/setting
6. A shared perception of team cohesiveness

These properties of team cognition were present in the game play of Pandemic while the game environment allowed players to fail with little consequence and in so doing, reflect on how their team contributions impacted success or failure overall. For our analysis, we would note that it was the repeated trials across game play that provided the information "flow field" from which to detect the invariant structure of good distributed cognition. As our analysis suggests, most of the learning from Hadley's (2014) study of Pandemic came from players reflecting on game play in written memos. This was triangulated with the observations of game facilitators.

As stated, our view of game-based learning includes more than simply the interactions that unfold during the game itself. They include the metagame. Metagame interactions include post-game reflections and conversations, online searches for game-related hints, cheats, and tutorials, and apply to a series of game plays, as well as include creative reflection by players on how to formulate strategies for game play on subsequent occasions. It is the dynamics across game-to-game strategies and collaboration among game players outside any single instance of game play that interest us as much as the interactions within a single game experience or the final score from a single instance of game play. This perspective alters classical notions of assessment as starting points, midpoints, or endpoints as in the cases of placement (e.g., pretest), formative (e.g., midterm), and summative (e.g., final exam) evaluations. Instead, the entire set of temporal trajectories of coordinated activity between team players and across game trials becomes the means for defining operationally what playing a game means (e.g., Andrews et al., 2017; Kim, Almond, & Shute, 2016).

Dating back to classic educational games, like Oregon Trail, students are known to play games without "getting" the key educational premise (the learning objective). That is, some players adopt game strategies (goals and intentions) that are unintended by game designers. For instance, players can focus on "gaming" the game (e.g., accumulating achievement points or killing off cattle) or on goals related to other superficial features of play (accumulating gold in the World of Warcraft auction house), rather than on the play mechanics intended to align with curricular goals and objectives. For example, Caftori (1994) described how the competitiveness aspect of Oregon Trail led some players (middle schoolers) to race for the end of the trail as fast as possible, without regard for their companions or oxen. This defeated the main objective of the game designers to encourage social problem solving for survival and draw attention to difficulties of losses along the trail. In this case, the common video game attraction of killing for loot replaced the instructor-intended and designer-intended student learning outcomes from game play:

> The goal becomes so important that players neglect the health of other travelers and their own lives. Another example, shooting animals for food, was designed to teach children about different animals in different terrain, as well as be part of the reality of life on the trail. However, "shoot 'em up" has become a focus of attention

> for many students (mostly boys). Unfortunately, besides eye-hand coordination, not much else is learned. And eye-hand coordination is not one of the stated objectives of this game.
>
> **(Caftori, 1994, p. 6)**

This simple example highlights that game play is an emergent interaction that results from player strategies that arise from a dialectic interaction between game affordances and player intentions, on the fly. It is this property of board game play, essential to understanding games as assessments, that we focus on in this chapter.

As a final part of our introduction, we would like to separate board game play from typical action video game play. Board game play typically takes place on a much more relaxed and social scale than action video games. Recent work highlights the general cognitive impacts that highly stimulating action video games can have on overall cognition and of course distributed cognition, including verbal interactions as well as spatial skills and attention (Bediou et al., 2018). But for our purposes, we would like to focus on the impacts of interacting with others over a table-top board game where game mechanics draw from board positions, there is time for extended conversation and strategizing (turn taking rather than timed responses), and changing conditions related to random card selection and/or dice rolls. While engaging at a social level, we would suggest that the excitement levels and cognitive attentional demands of board game play would not parallel the impacts on general cognition of playing action video games. The cognitive impacts we observe tend to emerge less from overall cognitive arousal, but from dynamic social interactions, problem solving, and group dynamics.

THEORETICAL FRAMEWORK

The theoretical framework for our work draws from J. J. Gibson's (1986) ecological psychology of individual perception-action that defines human behavior as a dynamic interaction between individuals and their environments. An interaction is taken as the fundamental unit of behavior (unit of analysis), thus co-defined by properties of individuals that enable them to affect the world (effectivities) and properties of the environment that afford or invite actions (affordances). This differs from the predominant cognitivist representational view of cognition in several ways and these can be illustrated in an analysis of typical board game play.

For the purposes of our analysis, we next introduce a few key concepts:

- The information flow fields (visual, auditory, tactile) of game play
- Detection of invariance among a sea of variance within the information flow fields across games
- Affordances detected in the flow field that are co-determined by effectivities of the player and develop during game play
- The hierarchy of intentionality and associated ontological descent of intentions during a single game and across instances of the same game
- The boundary constraints created by goal adoption and rule sets of game spaces

In describing human perception and experience, the Greek philosopher Heraclitus of Ephesus is famously quoted as saying, "You can never step in the same river twice." Drawing from Gibson's ecological psychology we similarly would contend that you cannot play the same game the same way twice. While you might be guided to enact the same strategy twice, the changing information field created through interactions with other players, random events in the game such as dice rolls, and your own attention to certain details (and not others) mean the game experience as an act of perception-action that unfolds on the fly cannot happen identically twice. We suggest here that game play is thus an interaction between player goals and intention and the *information flow field* created by the actions of other players and unfolding game events. For us, rather than "turn" or "score" the basic unit of analysis in game play is a player-game position interaction within a context bounded by the game rules and the intentions adopted at the moment by the player/team.

These information flow fields that are continuously signaling problems and changing conditions that the goal-driven agent (i.e., player) can interact, detect, and act on provide the context in which each player/team takes action (or makes moves). In doing so, they are guided by their goals and intentions (for winning) that direct their attention guiding their "pick up" of information directly from the environment, particularly certain regularities in the information flow field that specify affordances for action. These affordances are detected as invariant structures, much as Gibson described consistencies in the visual flow field when observers walked around a stationary object like a cube. As one moves, some things, like the visual background, move while some visual relationships (internal to the object) remain invariant. Those invariant structures specify the object and its properties (such as its graspability). Similarly, we would suggest that there are invariant structures in the information flow fields of game interactions that specify affordances for individual player "moves" and strategies as well as emergent (new) collaborative group objectives that can be perceived and acted on.

An elaboration of Gibson's ideas was added to this analysis when Shaw and Turvey (1999) described the affordances that are detected in the information flow fields as *co-determined* by properties of the individual and properties of the environment, an agent-environment interaction. Properties of the individual include what cognitivists would call their skills and abilities, but what are more precisely described within ecological psychology as "effectivities" or an agent's abilities to have an effect on the environment through their actions. This would include their physical abilities to roll dice and move tokens around a board, but also their abilities to engage with other players in shared decision making and civil discourse and the skills (gear, weapons, etc.) their character acquires in the game.

To Gibson's ecological framework of cognition, we add the contemporary learning science of situated cognition that draws heavily on apprenticeship learning and describes learning as the movement of individual behaviors from peripheral participation within communities of practice toward central participation. Situated cognition (Brown, Collins, & Duguid, 1989; Lave & Wenger, 1991; Rogoff, 1990) described learning as inherently social and a process of moving from the novice status as a peripheral contributor to the community of gamers toward a more central role of expert player. The ideas of situated cognition date back to John Dewey (1938) and the insight that we learn by doing. This has a modern equivalent in Papert's (1991) constructionism (the word with the "N") builds on constructivism (the word with the "V") by adding that it is only by building (constructing) artifacts that we genuinely come to know things. We apply this framework to games and playful learning in general, and specifically for this chapter, to collaborative and competitive board games that may serve as assessments.

A SITUATED DESCRIPTION OF LEARNING FROM GAME PLAY

To learn or play a game is a goal-driven dynamic process of perceptual tuning (Gibson, 2000). Across a player's experience within a single game, and across several instances of game play, a player begins to detect invariance in the information flow fields (visual, auditory, and tactile) of the game that specify objects and concepts (including team concepts) that are part of game strategies, rules, movements, and outcomes. Outside the game, a player learns from discussions (online and face-to-face) with other players and eventually moves from status as a novice player to a more central contributor to the gamer world.

Fundamental to our description of game play is the idea that human behavior is a dynamic interaction of perception and action, a person-environment interaction. Action drives and defines what we perceive and perception guides our actions. Thus, perception and action co-define each other. The reciprocal nature of perception is often missing from a traditional description of behavior and from cognitive or behavioral assessments (Young, 1995; Young, Kulikowich, & Barab, 1997). Perception is often described as a passive cognitive act, by individuals and by teams. In describing board games, this would be the equivalent of assuming that the rules of the game are the rules that are guiding each player's actions. But for our purposes, a player's actions are determined on the fly, in each dynamic interaction with the game and other players. Thus, the chances that a player will play the game exactly the same on two occasions, or that a team will act identically from game to game are essentially impossible. That is not to say that regularities in the game and in the play of other players will not result in observable or testable consistencies in the overall interactions that constitute game play.

We have selected to focus on collaborative board games because, in addition to an individual player interaction with the game, there must also be a coordinated team interaction. While each individual typically has some unique skills defined by cards or dice rolls, there are also collective team goals that are only achieved through the coordinated action of all the players. While this adds a layer of complexity for a full description of game behavior, it also adds a potential source of information for assessment of team cognition in relation to individual achievements by creating a participatory environment (Gee & Hayes, 2012) where civil discourse, informal mentorship, and intentional learning can occur and be detected. That is, not only may collaborative board game play provide evidence of an individual player's leadership and communication skills, it may also externalize how a team discerns problems and takes coordinated actions. In short, collaborative games, in comparison to competitive play, represent a richer context for distributed team cognition.

During any particular instance of a game, players must respond to the current board position, their own situations determined by cards, rolls, and prior accomplishments, and also perceive the trajectory of the game as the situation unfolds with other players. We could identify three types of goals for collaborative board game play:

- Individual interactions with the board
- Team co-actions with the board
- Individual perceptions of teammate interactions and trajectories

Since in a collaborative board game, players must also coordinate their actions among their fellow players to work toward team objectives, we can distinguish goals the team shares and associated coordinated actions they take together from individual player's actions and turn taking. All levels of game learning are done on the fly and in the moment and can be described by the ecological psychology of Gibson's perception-action. This is, team goals are dynamic, as they are assumed to adjust with each individual play, and individual play is assumed to adjust to each changing game condition as well.

In addition to these three types of goals that unfold during game play itself, there are also game-related activities that occur outside of game play (meta-game interactions). Across instances of game play, players engage in reflective dialog with other players using online cheat/hint sites as well as live conversations about the game. This level of game learning seeks to pick up invariances across games to detect team interactions that proved successful or problematic, and to detect strategies that had worked in other games that had not yet been experienced directly through their own game interactions. Both within and across games, we would posit that the learning is well described as perceptual tuning.

To this we would add that within the community of gamers, there is a dynamic of social progress from peripheral players (passive readers of game sites) toward a more experienced central role (contributor to game sites) that helps define and possibly even develop the game environment. While many games are classics and evolve little through time (Monopoly, Scrabble) other games do change with user content (Cards Against Humanity, Trivial Pursuit).

GOALS AND INTENTIONS: INDIVIDUAL AND TEAM STRATEGIES

From the ecological psychology perspective, human behavior is an emergent interaction between an intentional agent and a dynamic environment. The agent-environment interaction is the basic unit of analysis for any understanding of game play or distributed cognitive teamwork. On any particular occasion of human behavior (a speech act or game move), our framework holds that the intentionality of the moment is itself a complex interaction among a hierarchy of competing goals—a hierarchical ontological descent of goals to the moment of a game move. By this, we are not simply referring to Maslow's 1943 hierarchy of goals, but to a complex hierarchy of intentions that each individual may adopt that exist and play out on a variety of space-time continua, including the low-level muscle movements and the highest-level game-long play strategies.

Many have argued that teams think. We agree that team cognition is more than the sum of individual thinking (perceptions and actions). Our ecological perspective on intentionality has been described in more detail in Young, DePalma, and Garrett (2002). Here we reframe that discussion away from butterflies and termite mounds (many simple animals' intentions to eat and reproduce, when taken together create a shared, collective intentionality and coordinated activity to maintain, expand, and work toward functional goals within an ecosystem), toward a focus on the specific environments of humans playing collaborative board games. Board game designers create constraints on player interactions by constructing game rules. Designers hope to limit the interactions among players and the board to a specific goal space that reduces the degrees of freedom in normal life interactions with the world to a much smaller set of "valid" game moves. This limiting of the problem space may be particularly valuable for using board games as assessments. To explain the majority of board game interactions, it is fair to assume that players/teams have adopted the goal to win the game, and thus their intentions all draw from the state space of legal game actions. It is worth noting that this would not always be the case, and sometimes players may adopt goals to make illegal moves or have a play goal to intentionally undermine or disrupt game play or team success (the betrayer mechanic, as built into games like Betrayal at House on the Hill). Presumably, these non-standard game interactions could be detected somewhat readily, during and as a result of play assessments in the form of inferred player short-term and game-long goals.

Collaborative game play can be viewed as values realizing (Zheng, Newgarden, & Young, 2012). Drawing from Hodges (2007, 2009), our work with language learning through collaborative play in World of Warcraft (five player raid parties) shows how various individual goals for play and for team-building can be braided together with team play. Thus, we view games as ecosystems where value-realizing dynamics come into play. Values are real goods, intentions of the system that can only be realized through perception-action within a particular event/context. From an ecological psychology perspective, language is a perception-action-caring system in which speaking and listening "demand an ongoing commitment to directing others and being directed by them to alter one's attention and action so that movements from lesser goods (i.e., one's present board position, achievement or goal) to greater goods (e.g., values) is realized" (Hodges, 2007, p. 599). As such, values are not only properties of a person but are also about relationships and the demands with all others in the context. Collaborative board game play is a socially constrained field where all game decisions and actions are constrained and legitimated by values of the ecosystem with all the others (e.g., other players, board position, game moves). That is, it is a jointly enacted field, intended to realize shared values (i.e., achieving board positions and other subgoals and ultimately winning the game).

Dixon, Holden, Mirman, and Stephen (2012) organized their manuscript around a central question that we would pose to all those interested in team cognition: "Does cognition arise from the activity of insular components or the multiplicative interactions among many nested structures?" We agree with Dixon et al. that there is mounting evidence from a wide range of domains suggesting that cognition is multi-fractal, emerging from dynamic in-situ interactions and distributed across teams, rather than stable latent traits or constructs dominated cognition (see also Hilpert & Marchand, 2018). Assessment models that break down human cognition into numerous independent components, such as memory, motivation, self-efficacy, interest, spatial skills, verbal skills, and reasoning, then apply an additive model and attribute all else to error variance, may well mask key distributed cognition dynamics that underlie team coordination and thinking. Perhaps a more positive takeaway from the multi-fractal analysis is that you can study cognition at any level (collaborative game play) and expect to see parallels at other levels (individual game play, applied collaborative teamwork).

Similarly, Cooke, Salas, Kiekel, and Bell (2004) argued that team cognition is more than the sum of the cognition of the individual team members. Instead, "team cognition emerges from the interplay of the individual cognition of each team member and team process behaviors" (p. 85). They argued that to measure team cognition, a holistic level of measures is required rather than collective metrics or aggregation procedures. The holistic measures focus on the dynamic team processes and co-actions taken by the team as a whole. In this regard, team consensus or communication consistency, team process behaviors, and team situation awareness (and associated taken-as-shared goals) are posited as potential measures of team cognition (perception-action).

Alternatively, we would like to begin to build an understanding of game interactions and their potential for assessment. Using our current learning games, we would like to find 100 or so examples and consider training a deep learning network like Amazon's AMI and let it analyze for determining features that might align with 21st-century skills of collaboration or instances of micro problem solving that occur during game play (in board games like Expertise), as they have been used in speech recognition, recognizing Chinese handwriting, zip codes, classifying text as spam, and computer vision. One approach would be to create labels of successful and unsuccessful collaborative game play activities, and we might perhaps proceed toward a deep learning network solution for recognizing those activities during game play.

To this approach we would want to add individual and team stated goals and objectives for play, discerned from retrospective game play analysis. This would enable us to begin to construct a description of game play as situated cognition and action. Toward this purpose we next describe one collaborative board game designed to enable and assess the team cognition of teachers working together to wisely integrate technology into classroom instruction.

PLAYFUL ASSESSMENT FOR EXTERNALIZING TEACHERS' TECHNOLOGY INTEGRATION SKILLS: THE BOARD GAME EXPERTISE

Board games are just one form of the broader category of playful learning. Board games feature board "position" of players or tokens and often involve cards or dice to introduce chance events. When board games are collaborative, there are aspects related to individual players' actions and decisions, as well as those related to the effectiveness of the team as a whole. Of course, when considering playful approaches to assessing distributed cognition, the traditional concerns of the assessment to provide evidence that is consistent (i.e., reliability) and credible/useful (i.e., validity) must be addressed. If an assessment was not able to validly capture what it is supposed to measure, it has little value formatively or summatively.

Traditional test theory relies on classical test theory, which assumes that observed scores are either simply the sum of true scores (i.e., true ability) plus error scores, or in the case of latent trait or item response theory (IRT), that logistic functions specify the relation between ability (e.g., achievement, reading comprehension, problem solving) and the probability of getting an item correct on a test. These classical psychometric frameworks assume that there is an individual factor and all else is error, taking no account of other factors such as context. However, in alignment with a perspective of situated cognition, we view that cognition is always situated in context of a world interaction, and contextual complexities (i.e., constraints) can substantially co-define (along with individual abilities) the nature of many activities. As such, our true ability always exists and operates as an interaction within the environment on the fly, co-defined by properties of the individual and properties of the environment. It means that the error term from classical test theory may not be entirely error at all, but rather needs to be taken into account to define our true ability (Bateson, 2000/1972; Hutchins, 2010; Young, 1995). We thus need to integrate contextual complexities into any analytic framework (i.e., unit of analysis) for the validity of an assessment and to examine to what extent the interaction with the assessment context fully represents what happens in the real world, as well as to what extent the assessment context is really related to an applied practical contexts (Wiggins, 1993), with the goal to produce evidence showing that the results from the assessment really capture one's authentic practice. Mislevy (2016) made these points in an article entitled, "How Developments in Psychology and Technology Challenge Validity Argumentation," and he drew specific attention to situative/socio-cognitive psychological perspectives as reasons for needed developments in psychometric modeling given opportunities to design interactive performance assessments with digital technologies. We will summarize some of these developments in this chapter. However, we first introduce the board game Expertise to anchor our explanation of the theoretical assessment premises.

To establish such an ecological validity or contextual fidelity, we implemented a collaborative board game called Expertise as an assessment environment and examined to what extent the play scores from this collaborative board game parallel real-world practice in regard to teacher's technology integration skills. The unique attributes of Expertise as a collaborative board game constrain, to some extent, the large degrees of freedom present in any social interaction involving as many as five players.

EXPERTISE: THE GAME

The game play we are investigating comes from a board game we created to assess the wise technology integration skills of master teachers, a game called Expertise. This game is the result of several iterations of game design. Two theoretical frameworks, our ecological and situated cognition framework, and the TPACK framework (i.e., teacher knowledge of technology integration), guided its development.

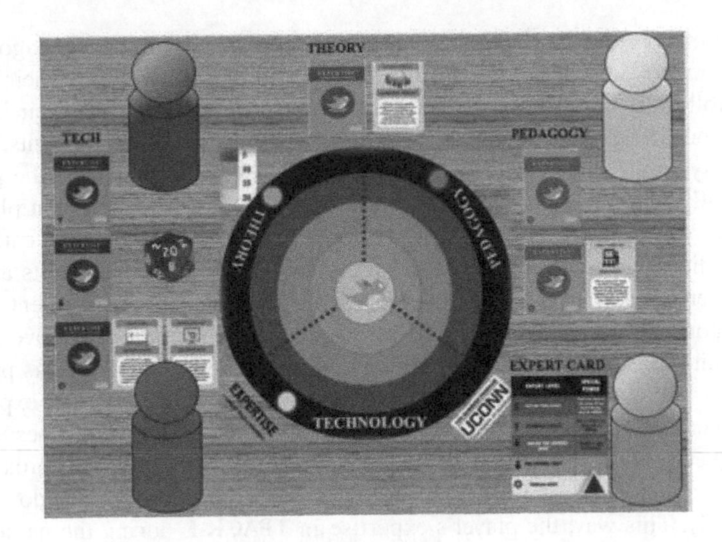

FIGURE 3.1 Expertise game initial game position.

The Expertise game set up is shown in Figure 3.1, including one board, three technology card decks, one theory card deck, and two pedagogy card decks. Each player also has an expert level card (tracking individual scores) in light of TPACK-L performance over the course of the game play. Up to five players can play this collaborative game that draws on team distributed knowledge and coordinated action. In every round, selection of two technology cards, one pedagogy card, and one theory card creates the environment or instructional context for player collaborative action. The game starts with one player (the speaker) sharing a curricular student learning objective with a content area for a lesson that could involve technology integration. In this first phase, the team serves as co-teachers or technology integration advisors to help the speaker construct possible technology integrations. In the second phase, each player judges a brief summary of the proposal presented by the speaker. The game runs two rounds, meaning all players rotate through the role of speaker two times and serve as advisors when not the speaker. In the first round, players use the first-round pedagogy card deck that includes more traditional instructional strategies such as direct instruction, group discussion and the like; in the second, more advanced round, they use the second-round pedagogy card deck that includes more complex and innovative contemporary teaching strategies, such as problem-based learning, gamification, anchored instruction, and the like.

Once the speaker shares the content area, all of the players together have three minutes to discuss their best collaborative solution to teach the given content by taking the given technology, pedagogy, and theory into account. After this co-construction process is done, the speaker has two minutes to state this solution (as if presenting it to a Board of Education for funding) and how the chosen technologies will be integrated with the given pedagogy and learning theory. During the speaker's proposal, each of the other players serves as a reviewer in one of the TPACK areas. Respectively, they decide to what degree the speaker's proposal addresses wise use of board *technologies*, sound *pedagogy* aligned with the pedagogy card, and alignment with the board card for *learning theory*. To structure their judging, players roll dice to indicate their role play predisposition regarding their judgment and are thus assigned a degree of harshness to apply to their judgments, ranging from "easygoing" through "critical" to "hard-ass" for each round. If the proposal successfully hits any of the components of TPACK-L, the game token placed for each of the TPACK-L components moves one step inward toward the center of the board, indicating the level of team success. Once co-players' judgments are done, the speaker's expert level is promoted based on the result of the judgment. If all of the components are successfully addressed (i.e., three components move in)—the speaker hits TPACK-L reasoning perfectly—his/her game expert level is promoted two ranks. If the proposal meets two components of TPACK-L, his/her expert level is promoted one rank. If the proposal meets only one component or does not meet any, his/her game level stays at the current level. But if the proposal fails to meet any of the components of TPACK-L components, his/her game level is downgraded accordingly. This way, the player's expertise in TPACK-L during the game play is given an individual score. In a nod toward scoring distributed cognition, after judgment is done, the speaker also has a chance to elevate the individual score of one of the co-players who was most helpful that round. There are five expert levels: Two Blue Agent, Two Summer Agent, Master Two Summer Agent, Technology Coach, and Master Technology Coordinator. At the end of two rounds of play, the final board (team) position and each individual player (expertise) level are recorded.

In addition, in each round, players are asked to use a technology card deck according to their expert level. Suppose that the player expert level is a Two Blue Agent, then this player should use the Two Blue Agent technology card deck. This way, game technologies become more sophisticated for some speakers as the game proceeds.

While beyond the scope of this chapter, the description of Expertise and its design helps to establish the types of data streams that are possible as means to study complex, interactive team performance and distributed cognition in game environments. The dissertation research of our third author provided evidence of the validity of Expertise as a measure of teacher technology competence. These preliminary research results point to the value of estimating both individual player and team success parameters in the spirit of an ecological description of game play.

PSYCHOMETRIC ISSUES OF GAMES VS. TESTS AS ASSESSMENTS

Because games are designed with rules and lead to individual and team outcomes that can be observed directly as win, lose, or draw, they provide a means for actions and behaviors to be analyzed as ones that result from goal-driven, intentional dynamics (e.g., McKee, Rappaport, Boker, Moskowitz, & Neale, 2018). One could argue that attempting to get items correct on any familiar standardized achievement test, whether its format is multiple-choice or constructed response, is similar, and learning can occur from completing tests and retests, just as it can from playing a game and replaying it. For example, in their now classic piece entitled "The Theoretical Status of Latent Variables," Borsboom, Mellenbergh, and van Heerden

(2003) considered how Albert Einstein might go about completing a general intelligence number series problem, the Fibonacci series problem, 1, 1, 2, 3, 5, 8 . . .? as a sequence of steps leading up to the construction of an answer rather than just its simple recording as 13. They wrote:

> Let us reconstruct the procedure. Einstein enters the testing situation, sits down, and takes a look at the test. He then perceives the item. This means that the bottom-up and top-down processes in his visual system generate a conscious perception of the task to be fulfilled; it happens to be a number series problem.
>
> **(p. 213)**

Borsboom et al. continued the description making reference to working memory and drawing on information from long-term memory (e.g., Einstein recognizes it is the Fibonacci series). While working memory and long-term memory resonate with information-processing theory accounts of problem solving (e.g., Anderson, Reder, & Simon, 1996; Vera & Simon, 1993) rather than those positioned as Gibsonian descriptions of perception-action cycles, the described sequence of Einstein's hypothetical steps as a test-taking narrative does make for an important illustration. It shows how a simple problem, and just one test item, commonly encountered on a large-scale standardized test is an in situ experience that is based on a sequence of goal states continuously created and annihilated through on-the-fly agent-environment interactions by test takers as they (in this case, Einstein hypothetically) interact with the test environment. Borsboom et al. continued their description of Einstein's test-taking behavior as well as that of another agent who assigns a score to the response Einstein records as his answer:

> Now he applies the rule and concludes that the next number must be 13. Einstein then goes through the various motoric processes that result in the appearance of the number 13 on the piece of paper, which is coded as 1 by the person hired to do the typing. Einstein now has a 1 in his response pattern, indicating that he gave a correct response to the item.
>
> **(p. 213)**

Of course, modern test theory (e.g., Embretson & Reise, 2000; Markus & Borsboom, 2013), associated with the study of the reliability and validity of scores on standardized achievement and intelligence tests, does not unravel the perception-action experience of Einstein or any other learner as Borsboom et al. described. Further, Borsboom et al. were not offering an ecological psychology view of problem solving as the alternative to cognitive and *in the mind* theoretical perspectives such as information-processing (e.g., Anderson et al., 1996) or connectionist (e.g., McClelland, 1988) accounts. Instead, the goal of Borsboom et al. was to highlight that within-individual variations over time in "achievement," "creativity," "decision making," "motivation," "problem-solving," etc., and treated as latent variables, drivers of behavior within the knower, have been under-addressed in psychometric modeling frameworks. Instead, latent variable modeling frameworks that undergird the broad spectrum of current psychometric and statistical techniques accepted as best-practice trade tools for establishing the reliability and validity of scores are best understood as between-subjects variation examined in at least one point in time but open to study of the between-subjects variation over multiple points in time.

As described previously, ecological psychologists work from a different world view focused on perception-action dynamics, and these perception-action cycles or interactions cannot be understood by anything other than the contribution of the individual agent interacting with her/his environment over time (e.g., Young, 1993; Young & Barab, 1999). Therefore, variation is of the within-subjects kind, and yet, there are other important sources of within-subjects, perhaps "within-contexts-unit" variation that must be coordinated in any study of game play. Specifically, and at minimum, these include the dynamics within one's team and associated dynamics of the environment that result in dynamics of individual and team intentions *real* spacetime.

There are similarities, but also, there are differences between game play and test taking as a means to study learning, whether learning is latent, somewhat latent, or not. As an initial step in outlining a psychometrics for games as assessments, the distinctions between game play and test taking can be compared and contrasted (see Table 3.1). Psychometric theory, both classical and contemporary, relies extensively

TABLE 3.1
Characteristics of Game Play and Test Taking

Characteristic	Game Play	Test Taking
Attribute	Performance (individual and team)	Achievement
Psychometric status	Behavior/manifest	Construct/latent
Unit of analysis	Perception-action coupling and cycles	Individual, classroom, school, district
Stimuli of assessment situations	Board features, opponents, team members (interactions)	Directions, items, options
Responses to assessment stimuli	Moves	Item selection or spoken/written response construction
Feedback to responses	Immediate	Delayed
Goal and intentions based on feedback	Level up, improve strategy, try to win	More correct responses
Degrees of freedom	Governed by rules of the game	Governed by directions of the test and sampling space of items
Score assignment	Binary (i.e., dichotomous)	Binary (i.e., dichotomous), polytomous
Dimensionality of scores	Locally dependent, emergent	Locally independent, unidimensional/multidimensional
Analysis of scores	Differential, nonlinear	Discrete, linear
Evaluation of the scores	"Making progress," "we can improve"	"Your score is at the 50th percentile or 75th percentile, etc. which means that . . ."
Consequences given, evaluation of scores	Rewards for progress	Penalties for lack of improvement

on a test-taking perspective. Only recently is psychometric theory embracing the value of understanding game play (e.g., Kim et al., 2016; Mislevy et al., 2014) as a means to analyze learning and problem solving. Yet, the study of game dynamics is a field by itself in the discipline of mathematics. Finally, game play is much better suited as an assessment of problem solving from an ecological account of learning than standardized test taking because the environment is constrained or designed in such a way to permit analysis of player-board as well as distributed team interactivity as it unfolds in spacetime (e.g., Stephen, Boncoddo, Magnuson, & Dixon, 2009). Such game play spacetimes can be conceived of as goal states of each play against the board individually nested within team goal states as they emerge and are then eliminated when achieved.

In the next section, we compare and contrast characteristics of game play and test taking. Both activities require agents to assess their progress in formal and informal learning settings. Our analysis of game play applies the ecological worldview of situational ecological dynamics and could be described as "behavioral, yet intentional" rather than "latent" given definitions and perspectives described by Markus and Borsboom in their *Frontiers of Test Validity Theory* (2013). Further, we are encouraged by contemporary developments in psychometric and statistical analysis on specification and test of nonlinear dynamical systems models (e.g., Helm, Ram, Cole, & Chow, 2016; McKee et al., 2018; Molenaar, 2014) at an individual time series level of analysis as well as co-integrated (e.g., Boker & Martin, 2018) with another player's trajectory of intentional dynamics. The individual and co-integrated time series may be the most fruitful avenue of research as to the scientific study of distributed cognition within game play. However, while the avenue appears a promising and worthwhile one to take, challenges are to be noted and may be difficult to overcome at the current time and for some time. These challenges relate to theory (Borsboom et al., 2003); game design as assessment design (Kim et al., 2016); data capture (e.g., Lindell, House, Gestring, & Wu, 2018); and mathematical/statistical modeling (e.g., Greenberg, 2014) as well as difficulties that arise given standard, accepted educational best practices in the use of test scores to diagnose and promote learning. Such forces of accepted practices include the best explanation of achievement/performance is the most simple one (e.g., Templin & Bradshaw, 2014) and that the test score is "my score" not "our score" (von Davier & Halpin, 2013).

Table 3.1 outlines characteristics of game play and test taking that were selected based on the way psychometricians and statisticians approach data analysis to understand human learning (e.g., Markus & Borsboom, 2013). First, consider the primary attribute of game play versus test taking, which is an assessible variable, an entity that belongs to the learner and for which he/she provides data for measurement, assessment, and evaluation of assessment. For game play, we identify the primary attribute as "performance," which occurs in real spacetime and may or may not be influenced by prior experience. For example, consider the student who is playing a game like Expertise for the first time. At minimum, rules need to be learned. However, no one would penalize the player for her/his willingness to do better the next time. Comparatively, the primary attribute for taking a test is demonstrating "achievement" or perhaps aptitude or intelligence, which is after the fact and the result of rather than a feature of learning, "shaped by curriculum standards," and prone to evaluation at a level beyond any individual student (e.g., the teacher taught well, the school district operated well).

Next, we consider the psychometric status of the primary assessment attribute. The psychometric status pertains to how the reliability and validity of scores are to be interpreted as behavioral/manifest or formative versus latent and reflective. For game play, performance is behavioral as goals must be realized in action (e.g., Kulikowich & Young, 2001; Young, 1993). However, an important distinction is to be made between behavior that is reactive and responsive to the environment as in premises laid out by Skinner (1988) compared to intention-given by the learner. Gibsonian dynamics are intentional dynamics, and therefore, learners are often referred to as "agents," referencing their agency to shape experience given their continuous interaction with the environment. While some tests adapt to user responses, a game is much more dynamic with game moves and strategies taken as a given part of the context. In contrast, achievement in test taking is primarily treated as latent. "Reflective" is a psychometric term (e.g., Markus & Borsboom, 2013) and it means that any item or task response that is assigned a score depends upon or is regressed upon an underlying long-term stable latent trait that is not directly observed. As such, the latent trait, abstracted from the world and stored in the head of the users, is the cause of student responses. Treatment of attributes as latent traits is arguably the most popular psychometric approach to the study of scores as reliable and valid and is evidenced in the variety of confirmatory factor analysis (CFA) and item response theory (IRT) models that psychometricians specify and test. It is also important to note that these approaches are successful between-subjects covariation techniques where scores are compared and positioned relative to one another, most often in reference to location in a normal distribution—all such assumptions are not taken as given in an ecological psychology description of game play.

From the discussion of the psychometric status of the attributes of game play and test taking, respectively, it is hopefully becoming clear that the unit of analysis for game play must be a continuous cycle of perception-action couplings in a dynamic agent-environment interaction. These perception-action couplings are manifest given the effectivities of the learner with the affordances available in the game (learning) environment, and neither student nor environment remain the same. The process unfolds and presents a response stream that can be modeled statistically as situated experience. Therefore, there is no abstraction from the world or need for storage of ideas. Comparatively, the unit of analysis for test taking is the static student whose achievement is nested in the static contexts of higher-order or multilevel units of analysis—classrooms, schools, school districts, provinces/states, countries (e.g., Raudenbush & Bryk, 2002). Indeed, covariates (e.g., classroom enrollment size, school socio-economic status) can be entered into the hierarchical equations at any level signifying contextual elements that explain sources of variation in students' scores. Further, cross-level interactions (e.g., students' prior achievement scores crossed with school socio-economic status) can be evaluated for significance and effect size. However, these covariates or covariate interactions are not affordances students detect and with which they interact during learning. Further, for both students and for higher levels of analysis such as classrooms, variables are most often treated as trait-like and stable, not state-like and continually changing (e.g., Molenaar, 2004).

The next five characteristics in Table 3.1 address the data sources that researchers can analyze for both game play and test taking. These include (1) stimuli of assessment situation; (2) responses to assessment stimuli; (3) feedback given responses; (4), goals or intentions based on feedback provided; and (5) degrees of freedom for any agent to alter the course of game play or test taking based on any feedback provided. We unpack each of these briefly next.

The stimuli of assessment situations are arguably infinite and would include any sensory information that can be detected (e.g., light, sound, touch). However, as an assessment activity, these stimuli are select features that can be incorporated into the quantitative models to indicate students' progress. As such, they are part of assessment design. For game play, the stimuli would include board features (see Figure 3.1) as well as any additional metagame resources (e.g., notes, internet support sites) that allow the game activity to unfold on a single instance or across instances. Team members and opponents would also be part of the stimulus field. By comparison, while students can complete test-type exercises collaboratively (e.g., Borge & White, 2016), most often, and in the case of standardized achievement tests, they do so alone with limited stimuli that include directions, items, options, calculators, and notepads. All stimuli can be coded as contributing to classical item characteristics such as level of difficulty, degree to which the item distinguishes between low and high scores (i.e., item discrimination), and guessing. For example, Gorin and Embretson (2006) have used this strategy to understand more about students' answers on paragraph comprehension items beyond basic estimates of a one- or two-parameter logistic IRT model. Often called multicomponent latent trait models (Embretson, 1985), these IRT models are informed by cognitive theory and are applied best when tenets of the theory are incorporated into the assessment design. Gorin and Embretson (2006) described construction of a spatial analogical reasoning task, like Raven's Progressive Matrices, and showed how addition, deletion, rotation, and transformation of item stimuli correlate with item difficulty and estimated ability or proficiency of examinees. Still, this approach to assessment, data capture, and psychometric modeling is more trait-based, stable, and between-subjects than what can be modeled for game play, and this leads to the next characteristic of Table 3.1, and potentially, one of the most important: the manifest response indicators (Markus & Borsboom, 2013) that become the data streams by which psychometricians and statisticians test their models.

The primary responses for game play are the moves or the turns that each player takes and contributions they make to team planning and distributed cognition. Most games require several moves or turns in order to arrive at the final outcome: win, lose, or draw. Consider three of the most classic of all games: checkers, chess, and tic-tac-toe. Like Expertise, these are board games, but simpler in design. All have multiple stimuli. All require multiple moves, often timed moves as in the case of elite-performance chess, and all progress in such a way that "on-the-board" stimuli (e.g., game locations, pieces) decrease in time. For test taking, there are some similarities; however, there are also key differences. Test items do afford a sequence of responses as either selected on multiple-choice tests or constructed for short answer or essay questions. Further, any situation of test taking can be timed as in the case of the Scholastic Achievement Test (SAT) or Graduate Record Examination (GRE). However, the student who is likely attempting to provide her/his best performance, response after response, etc., whether it is selected or constructed, is unaware of what the test developers' intentions and goal dynamics are as each item is encountered. This lack of information also applies for computer-administered tests as in the case of computer adaptive testing (CAT) where items are tailored to examinees as they record or submit each response in sequence (e.g., Lord, 1968; Van der Linden & Glas, 2000). As such, there is little feedback provided to the examinee, certainly no knowledge of results, and not even guidance in a form such as, "Here is an easier item. Try it." or, "Here is a more difficult item. I think you can do it." Instead, the test taking endeavor is to maintain the calibration of items as positioned given a between-subjects' evaluation as below average, average, or above average within most likely a normal distribution. Further, environments for test taking often require a "standard setting," which means that the conditions (e.g., proctor, seating arrangements, time limits) for completing the assessment remain constant for all examinees as possible.

Next in Table 3.1 the characteristic feedback to responses highlights that feedback for each move or turn taken in game play is immediate, and that immediate feedback resets the dynamics of intentions (a single move) to intentional dynamics (e.g., how has my individual strategy or my team's goals changed? (Young & Barab, 1999)) unless play is disrupted or interrupted as with the end of the game or game delays as in the case of outdoor games (e.g., weather delays). By comparison, feedback to responses that students provide on tests can be significantly delayed. While contemporary standardized assessments, like the SAT and GRE, provide computerized score reports upon completion of the tests, the examinees are not provided any feedback while taking the test. In other test taking situations such as classroom assessments, students might wait days or weeks before they have a sense of their progress. Finally, as in tests administered for research purposes, participants are unlikely to receive any feedback about how well they performed unless they were to inquire about their results at the conclusion of the study.

These points made about the immediacy or the delay of feedback relate directly to the next characteristic in Table 3.1. While Gibsonian accounts of perception-action are as relevant for game play as they are for test taking, as we have written previously, game play affords more opportunity for discussion and adoption of goals related directly to learning to improve performance as part of the activity (e.g., preparing for the challenges at the next level; altering strategies to improve; identifying opportunities for distributed cognition among one's teammates) than does test taking, which in some ways can only mean attempting to get as many or more correct answers on a test, even if guessing. The overt learning goals of game play highlight how players may be operating at multiple levels of an ontological hierarchy of goals, and within particular instances of games, may not appear to be showing optimal performance, as in the case of testing out strategies or stretching the rules of the game (e.g., playing "what will happen if I try this?").

These goal reformulation properties are related to what can be called the degrees of freedom for each of the two measurement situations. In statistics, degrees of freedom are values that are free to vary. While the term primarily pertains to samples selected from populations in the estimation of parameters, the idea is relevant for our current discussion. Degrees of freedom relate to the number of parameters that constrain one's perceiving and acting. In effect, they relate to constraining the opportunities for action. In some situations such a free play on a school yard, as with most human situations, there are nearly infinite degrees of freedom (Mitchell, 2009). In situations like the card game War, there are fewer degrees of freedom and performance can seem repetitive, leaving little behavior to assess. In most games, the rules are set at the start of play, but in some games (like Flux) the game rules themselves emerge from play or are initially hidden to all but a few players (as in Betrayal at House on the Hill). In these cases, the degrees of freedom are changeable and must be assessed continuously throughout play. Whether emergent or not, the rules of the game establish the degrees of freedom for game play. The patterns of turn-taking, rolling dice, drawing cards, and discussing next moves with teammates are part of the rules of the game.

Note the use of gerunds in the preceding sentence. These nouns derived from verbs illustrate not only the value of action, but also, each gerund defines constraints under which game play can unfold. For test taking, it is no different. Test taking involves following directions, reading stems, selecting among options A to D, and constructing short answer responses or short essays as in the case of the National Assessment of Educational Progress (NAEP). Consider two hypothetical NAEP-like items, one is multiple choice and one is short constructed response.

Multiple-choice item: Which of the following psychologists focused his work on the importance of visual perception?

A. Gibson
B. Piaget
C. Skinner
D. Vygotsky

Short constructed response item: For Skinner, stimulus-response connections defined schedules of reinforcement. In the space provided below, define "stimulus" and define "response" according to Skinner.

There are several important points to consider about such items. First, the actions of examinees are arguably more passive or reactive than those for game play. Second, except for the responses that examinees can construct, the constraints are so limiting (e.g., finite option set, restricted space allocation for writing a constructed response) that there is limited opportunity for students to interact with the environment in such ways where their responses now provide affordances for further action by an evaluator, team member, teacher, etc. So, in essence, the flow of activity has stopped with the response for that item stimulus field. Examinees must then *re-situate* to move onto the next stimulus field (i.e., the next item on the test) that may present content that is significantly different from the item just completed.

We acknowledge that many computer-based assessments are now designed to relax some constraints (e.g., Jang et al., 2017; Siddiq, Gochyyev, & Wilson, 2017). For example, Quellmalz and colleagues (2013) studied scores of three different "next-generation" assessment environments designed in accordance with school day standards such as those of College Board Standards for Science Success (College Board, 2009) and the National Research Council's Framework for K-12 Science Education (National Research Council [NRC], 2012). Referred to as "static," "active," and "interactive" designs, Quellmalz and her research team demonstrated psychometrically that as environments became least constrained (i.e., interactive), the dimensionality of the responses (i.e., how many factors or latent traits predict performance) increased.

Static modality item designs looked very much like the two hypothetical examples we presented previously. As one example, students looked at four different ecosystem food chain diagrams as a stem that depicted relationships among animals (i.e., bear, caribou, and hare) and plants (i.e., grass and lichen). Then students read a description of a food web diagram. They had to select the correct option that indicated a match of description to diagram.

In the *active* and *interactive modalities*, students' opportunities to use menu tools to construct food web diagrams increased. For example, students could view simulated animations to watch predator-prey ecosystem dynamics. With the ability to control viewing and re-viewing of the videos, the active modality afforded examinees more control over managing information before they would either select a response among multiple-choice options or construct a response using arrows to diagram food web flow. Finally, the *interactive* modality permitted the highest degree of activity (and degrees of freedom) and authentic engagement for students. In this modality, learners could conduct scientific inquiry using data streams much as expert scientists do before they would select or construct an item response. Much as in game play, the interactive modality updated information flow based on each input or "move" made by the student.

Tracking the data via computer logs (e.g., Tsai, 2018) allows for process tracing (Lindell et al., 2018) that is an entire specialization in design of technology-rich learning environments (Jang et al., 2017) used for the study of dynamic decision making. Large amounts of data (big data) can be harvested and used for information systems research (Barki, Titah, & Boffo, 2007). This set of topics also leads to the next characteristic in Table 3.1, score assignment. For highly interactive modalities such as game play, any move, such as a keystroke, menu review, roll of dice, or selection of a card from a deck can be coded dichotomously as present or absent (i.e., binary 1 or 0) as well as time stamped for duration and even position (space coordinates). By comparison, test taking, described as a static assessment modality, can lead to score assignment that is either dichotomous (e.g., disagree/agree, incorrect/correct) or polytomous (e.g., Likert scale, partial credit). In fact, IRT models vary given the type of assessment (e.g., achievement measure, affective scale) as well as the score assignment (i.e., dichotomous or polytomous). Masters (1982) proposed a partial credit Rasch model to scale distractors on multiple-choice items given degrees of correctness. Similarly, Andrich (1978) presented a Rasch rating scale model to evaluate the contributions of Likert-scale categories (e.g., strongly disagree to strongly agree) in score reliability and validity. Numerous other IRT models have been introduced in psychometrics literature for polytomous score assignment that might prove useful for game play assessments (Embretson & Reise, 2000).

What the vast majority of IRT models share in common leads to the next topic of Table 3.1, dimensionality of scores, which pertains to a topic in statistics called *local independence.* for test taking, the only property connecting the responses as a sequence is the one or more latent traits that predict the responses as manifest item indicators (Markus & Borsboom, 2013). Various models of dimensionality exist including the classical one-, two-, and three-parameter dichotomous IRT models (Hambleton, Swaminathan, & Rogers, 1991) and their extensions to polytomous singular and multiple dimension IRT models (Reckase, 2009), to bifactor models (DeMars, 2013), and even to multilevel IRT models (e.g., Fox, 2004; Wilson, Gochyyev, & Scalise, 2017). In these more advanced models, estimations of item parameters are studied given variance contributed by clusters (e.g., dyads, groups, classrooms, schools, districts) as well as individual, which potentially could capture both player and distributed team cognition. Maximum likelihood estimation facilitates the estimation of parameters and study of model-data fit so that error or residual variance is minimized given score patterns. Therefore, when model-data fit is supported, then scores can be summed as a total composite if unidimensional and as discrete subscales if multidimensional (e.g., Dumas & Alexander, 2016; Schilling, 2007). This can facilitate statistical analysis using general linear model (GLM) procedures and their extensions (e.g., HLMs, HGLMs) to address classical research questions such as, "Are there significant differences in instructional treatment conditions on reading comprehension after controlling for prior knowledge?" (e.g., McNamara & Kendeou, 2017), or "Do offline and online reading skills predict critical evaluation of Internet sources?" (e.g., Forzani, 2018). However, these group-based and aggregate modeling techniques (Boker & Martin, 2018) impede, and potentially prohibit, data-analytic science of learning and problem solving as it occurs most naturally for any one learner attempting to coordinate activity with any other learner, as in the instance of the distributed cognition of collaborative game play.

Game play, in contrast to test taking, cannot be anything other than emergent and nonlinear. Local dependence (see Table 3.1) is inevitable, and dimensionality, singular or multiple, unfolds within multiple nested spacetimes (e.g., McKee et al., 2018). As we introduced previously, there is no presumption of a latent abstraction underlying individual and distributed cognition. Summed scores for composite variables such as those of the many verbal (e.g., vocabulary) and performance (e.g., working memory) subscales of intelligence tests are not the measured attributes of interest. Instead, researchers use terminology of "complexity" (e.g., Hilpert & Marchand, 2018) and "dynamical systems" (e.g., Kringelbach, McIntosh, Ritter, Jirsa, & Deco, 2015) and dynamic concepts including "adaptation" (e.g., Mitchell, 2009), "diffusion" (e.g., Dixon et al., 2012), "emergence" (e.g., Hilpert & Marchand, 2018), "entropy" (e.g., Stephen et al., 2009), and "self-organization" (e.g., Greenberg, 2014) to describe and model phenomena scientifically. These are the dynamics that underly an ecological psychology description of board game play, and that we propose to capture in Expertise play as an assessment.

Analysis of assessment data relies extensively on a mathematics that is suited to such terminology for either game play or test taking. Historically, complex and dynamical systems rely on difference or differential equations (e.g., Dixon et al., 2012; Stephen et al., 2009). However, there are numerous challenges when undertaking mathematical and statistical modeling of such kind. As a primary challenge, only the simplest of models are tractable or have approximate solutions (Ocone, Millar, & Sanguinetti, 2013). Relatedly, modeling with differential equations makes assumptions about initial conditions. While it is nearly impossible for researchers to specify initial conditions of any complex social system such as learning and problem solving as it takes place in classrooms, the initial board and player conditions of games may be more tractable. A final limitation of use of such models is that their properties are not easy to understand mathematically or as results that can inform best practices or classroom decision making. As Hilpert and Marchand (2018) discussed in their review of complex systems for educational psychology, dominant component linear models of studying test scores (i.e., test taking) such as regression and their extensions to path and structural equation models (SEMs) will prevail, at least for the near future. These linear (static) models are entrenched in classical texts and manuscripts adopted by the field (e.g., Bollen & Long, 1992; Byrne, 2001). In their final remarks, the authors write:

> Integrating CS (complex systems) research into educational psychology may require more flexible thinking about research methods, particularly with regard to significance testing, commensurate forms of data, and generally what counts as sound evidence within empirical research.
>
> **(p. 15)**

However, such progress also reflects the final two characteristics we present in Table 3.1. For game play, considering emergent dynamics allows for evaluations of progress and improvement and rewards for improvement as students' goals become more numerous, their challenges increase, and distributed team cognition becomes more effective. This evaluation and rewards perspective often differs from typical course grades and high-stakes test scores that focus on percentile ranks, school comparisons, and possible penalties for lack of improvement. Substantial challenges for applying these complex dynamical systems analyses remain, including:

1. How to get away from the general linear model and take as fundamental nonlinear dynamics
2. How to characterize the multiple levels of intentions (for play) that may be factors in individual player game acts, and the shared coordinated action of teams
3. Effects of context (playing a game in a classroom may impose different constraints than playing it at home, for homework, with friends, with strangers, etc.)
4. How to assess individual play and team play on multiple simultaneous spacetimes defined by various conflicting and coordinated goals

Potential of Games to Assess Individual and Distributed Team Cognition

Games have long been attractive as learning environments given that games provide goals (i.e., objectives), rules for constrained interactions, relevant/immediate feedback, content mastery, and more importantly, create playful spaces where players are welcome to explore the context (i.e., engaged participation), share, and build the experience with others (i.e., co-construction and distributed team cognition) (Gee, 2003; Shaffer, 2006; Squire, 2006; Young, 2004). In a 2012 meta-review of the value of video games for classroom learning, Young et al. (2012) contended that individual game play is best described as situated cognition, while collaborative team play can be described as emergent effectivities of a collective of agents that comes together to act in synchrony on shared goals. Several additional studies have pointed to the affordances of games as useful assessment tools (e.g., Chin, Dukes, & Gamson, 2009; Loh, 2012; Shute & Ke, 2012). Games may create good assessment contexts that capture the interactive/situative nature of group and individual cognition and action (Schwartz & Arena, 2013; Shaffer & Gee, 2012; Steinkuehler & Squire, 2014; Young et al., 2012). For example, Shaffer and Gee (2012) argued that games are good assessments in that every action/decision that players make in a gaming context draws on players' abilities in the moment on many dimensions, and collaborative play hinges on the situated and embodied nature of team cognition and action (e.g., thinking in situ in the gaming context, strategizing while acting in the context). Steinkuehler and Squire (2014) also asserted that a game itself is a good assessment tool in that games enable us to track and monitor players' performance, provide just-in-time feedback on any performance, and offer rich data on problem solving in situ. In addition, some games enable us to make observations in authentic contexts where we create complex, realistic scenarios required to evaluate players' situative actions and cognition (DiCerbo, 2014).

The dynamics of collaborative board game play seem to parallel the dynamics of complex educational settings like public schools, and we have preliminary data showing positive correlations between games scores, specifically Expertise and other measures of master teachers' performance in the realm of technology integration at the graduate school level. Some of the interesting aspects of board game play as a form of distributed team cognition are that there is a defined pattern of turn-taking, so each individual has roughly equal opportunities to contribute to the overall team success within the constraints of their various roles. This again nicely constrains some of the degrees of freedom in group collaborative interactions, perhaps enabling complex systems modeling.

Potential of Games to Assess Collaborative Distributed Team Cognition

It seems clear to us that current assessments that rely on linear assessment models to evaluate collaborative cognition in game play bring with them limiting factors that strain our theory of human cognition. We recommend the adoption of a situated approach that may rely more heavily on big data computer processing and analysis of deep learning networks. There is no guarantee that these approaches will necessarily be better able to characterize the emergent interactions of individual team cognition as described by ecological psychology and situated learning. But we have to try. The playful learning that occurs in structured board game play may prove a sufficiently bounded context in which to explore the application of these approaches while keeping the degrees of freedom manageable.

We are encouraged by recent developments in psychometrics where attention has been given to modeling complex dynamics (e.g., von Davier, 2017) that could be specifically applied to distributed team cognition and collaborative game play. In a 2017 special issue of the *Journal of Educational Measurement* with guest editor Alina von Davier, initial presentations of network models, time series incorporated into IRT modeling, and multilevel multidimensional modeling individual student variation coupled with their peers (team player distributed cognition) provided evidence of promising developments that bring together learning theory, game design, big data capture, and psychometric analysis. In our future work, we look to contribute to this dialogue with analysis of the many unfolding data streams available through study of board games like Expertise. We also acknowledge there is much work yet to do to fully describe the dynamics that unfold on multi-fractal levels when individuals interact with game mechanics as individuals and collaborative teams. Fortunately, the tools and supporting theoretical frameworks appear to be emerging to support such research efforts. Our suggestion is that this work be framed within the ecological psychology world view and investigate individual and team cognition as "situated" and emergent in a dynamic agent-environment interaction that is organized by individual and shared goals.

REFERENCES

Anderson, J. R., Reder, I. M., & Simon, H. A. (1996). Situated learning and education. *Educational Researcher, 25*(4), 5–11.

Andrews, J. J., Kerr, D., Mislevy, R. J., von Davier, A., Hao, J., & Liu, L. (2017). Modeling collaborative interaction patterns in a simulation-based task. *Journal of Educational Measurement, 54*(1), 54–69.

Andrich, D. (1978). A rating formulation for ordered response categories. *Psychometrika, 43*, 561–573.

Arachchilage, N. G. A., & Love, S. (2013). A game design framework for avoiding phishing attacks. *Computers in Human Behavior, 29*(3), 706–714.

Barki, H., Titah, R., & Boffo, C. (2007). Information system use-related activity. An expanded behavioral conceptualization of individual-level information system use. *Information Systems Use, 18*(2), 173–192.

Bateson, G. (2000/1972). *Steps to an ecology of mind: Collected essays in anthropology, psychiatry, evolution, and epistemology*. Chicago, IL: University of Chicago Press.

Bediou, B., Adams, D. M., Mayer, R. E., Tipton, E., Green, C. S., & Bavelier, D. (2018). Meta-analysis of action video game impact on perceptual, attentional, and cognitive skills. *Psychological Bulletin, 144*(1), 77–110.

Bedwell, W. L., Pavlas, D., Heyne, K., Lazzara, E. H., & Salas, E. (2012). Toward a taxonomy linking game attributes to learning: An empirical study. *Simulation & Gaming, 43*, 729–760. https://doi.org/10.1177/1046878112439444

Bodrova, E., Germeroth, C., & Leong, D. J. (2013). Play and self-regulation: Lessons from Vygotsky. *American Journal of Play, 6*(1), 111–123.

Boker, S. M., & Martin, M. (2018). A conversation between theory, methods, and data. *Multivariate Behavioral Research, 53*(6), 806–819. https://doi.org/10.1080/00273171.2018.1437017

Bollen, K. A., & Long, J. S. (1992). Tests for structural equation models: Introduction. *Sociological Methods & Research, 21*(2), 123–131.

Borge, M., & White, B. (2016). Toward the development of socio-metacognitive expertise: An approach to developing collaborative competence. *Cognition and Instruction*, *34*(4), 323–360.

Borsboom, D., Mellenbergh, G. J., & van Heerden, J. (2003). The theoretical status of latent variables. *Psychological Bulletin*, *110*(2), 203–219.

Brown, J. S., Collins, A., & Duguid, P. (1989). Situated cognition and the culture of learning. *Educational Researcher*, *18*(1), 32–42.

Byrne, B. M. (2001). Structural equation modeling with AMOS, EQS, and LISREL: Comparative approaches to testing for the factorial validity of a measuring instrument. *International Journal of Testing*, *1*(1), 55–86.

Caftori, N. (1994). Educational effectiveness of computer software. *T.H.E. Journal*, *22*(1) 62–65. Retrieved from https://thejournal.com/articles/1994/08/01/educational-effectiveness-of-computer-software.aspx

Chin, J., Dukes, R., & Gamson, W. (2009). Assessment in simulation and gaming a review of the last 40 years. *Simulation & Gaming*, *40*(4), 553–568.

Clark, D. B., Tanner-Smith, E. E., & Killingsworth, S. S. (2015). Digital games, design, and learning: A systematic review and meta-analysis. *Review of Educational Research*, *86*(1). https://doi.org/10.3102/0034654315582065

College Board. (2009). *Science: College Boards standards for college success*. Retrieved from http://professionals.collegeboard.com/profdownload/cbscs-science-standards-2009.pdf

Cooke, N. J., Salas, E., Kiekel, P. A., & Bell, B. (2004). Advances in measuring team cognition. In E. Salas & S. M. Fiore (Eds.), *Team cognition: Understanding the factors that drive process and performance* (pp. 83–106). Washington, DC: American Psychological Association.

Crichton, M. T. (2009). Improving team effectiveness using tactical decision games. *Safety Science*, *47*(3), 330–336.

DeMars, C. E. (2013). A tutorial on interpreting bifactor model scores. *International Journal of Testing*, *13*(4), 354–378.

Dewey, J. (1938). *Experience & education*. New York: Kappa Delta Pi.

DiCerbo, K. E. (2014). Game-based assessment of persistence. *Educational Technology & Society*, *17*(1), 17–28.

Dixon, J. A., Holden, J. G., Mirman, D., & Stephen, D. G. (2012). Multifractal dynamics in the emergence of cognitive structure. *Topics in Cognitive Science*, *4*, 51–62. https://doi.org/10.1111/j.1756-8765.2011.01162.x

Dumas, D., & Alexander, P. A. (2016). Calibration of the test of relational reasoning. *Psychological Assessment*, *28*(10), 1303.

Embretson, S. E. (1985). Multicomponent latent trait models for test design. *Test Design: Developments in Psychology and Psychometrics*, 195–218.

Embretson, S. E., & Reise, S. P. (2000). *Item response theory for psychologists*. Mahwah, NJ: Lawrence Erlbaum Associates, Inc., Publishers.

Forzani, E. (2018). How well can students evaluate online science information? Contributions of prior knowledge, gender, socioeconomic status, and offline reading Ability. *Reading Research Quarterly*, *53*(4), 385–390.

Fox, J. P. (2004). Applications of multilevel IRT modeling. *School Effectiveness and School Improvement*, *15*(3–4), 261–280.

Gee, J. P. (2003). What video games have to teach us about learning and literacy. *Computers in Entertainment (CIE)*, *1*(1), 20–20.

Gee, J. P., & Hayes, E. (2012). Nurturing affinity spaces and game-based learning. In C. Steinkuehler, K. Squire, & S. Barab (Eds.), *Games, learning, and society: Learning and meaning in the digital age* (pp. 129–153). New York: Cambridge University Press.

Gibson, E. J. (2000). Perceptual learning in development: Some basic concepts. *Ecological Psychology*, *12*(4), 295–302.

Gibson, J. J. (1986). *The ecological approach to visual perception*. Hillsdale, NJ: Erlbaum.

Gorin, J. S., & Embretson, S. E. (2006). Item difficulty modeling of paragraph comprehension items. *Applied Psychological Measurement*, *30*(5), 394–411.

Greenberg, G. (2014). How new ideas in physics and biology influence developmental science. *Research in Human Development*, *11*(1), 5–21.

Hadley, K. R. (2014). Teaching teamwork skills through alignment of features within a commercial board game. *International Journal of Engineering Education*, *6*(A), 1376–1394.

Hambleton, R. K., Swaminathan, H., & Rogers, H. J. (1991). *Fundamentals of item response theory*. Newbury Park, CA: Sage.

Helm, J. L., Ram, N., Cole, P. M., & Chow, S. M. (2016). Modeling self-regulation as a process using a multiple time-scale multiphase latent basis growth model. *Structural Equation Modeling: A Multidisciplinary Journal*, *23*(5), 635–648.

Hilpert, J. C., & Marchand, G. C. (2018). Complex systems research in educational psychology: Aligning theory and method. *Educational Psychologist*, *53*(3), 185–202.

Hirsch, P. L., & McKenna, A. F. (2008). Using reflection to promote teamwork understanding in engineering design education. *International Journal of Engineering Education*, *24*(2), 377–385.

Hodges, B. H. (2007). Good prospects: Ecological and social perspectives on conforming, creating, and caring in conversation. *Language Sciences*, *29*, 584–604.

Hodges, B. H. (2009). Ecological pragmatics: Values, dialogical arrays, complexity and caring. *Pragmatics & Cognition*, *17*(3), 628–652.

Howard-Jones, P., Taylor, J., & Sutton, L. (2002). The effect of play on the creativity of young children during subsequent activity. *Early Child Development and Care*, *172*(4), 323–328.

Huang, A., & Levinson, D. (2012). To game or not to game teaching transportation planning with board games. *Transportation Research Record*, *2307*, 141–149.

Hutchins, E. (2010). Cognitive ecology. *Topics in Cognitive Science*, *2*(4), 705–715.

Jang, E. E., Lajoie, S. P., Wagner, M., Xu, Z., Poitras, E., & Naismith, L. (2017). Person-oriented approaches to profiling learners in technology-rich learning environments for ecological learner modeling. *Journal of Educational Computing Research*, *55*(4), 552–597.

Kim, Y. J., Almond, R. G., & Shute, V. J. (2016). Applying evidence-centered design for the development of game-based assessments in physics playground. *International Journal of Testing*, *16*(2), 142–163.

Koehler, M. J., Arnold, B., Greenhalgh, S. P., & Boltz, L. O. (2017). A taxonomy approach to studying how gamers review games. *Simulation & Gaming*, *48*(3), 363–380.

Kringelbach, M. L., McIntosh, A. R., Ritter, P., Jirsa, V. K., & Deco, G. (2015). The rediscovery of slowness: Exploring the timing of cognition. *Trends in Cognitive Sciences*, *19*(10), 616–628.

Kulikowich, J. M., & Young, M. F. (2001). Locating an ecological psychology methodology for situated action. *Journal of the Learning Sciences*, *10*(1 & 2), 165–202.

Lave, J., & Wenger, E. (1991). *Situated learning: Legitimate peripheral participation*. Cambridge: Cambridge University Press.

Lillard, A. S. (2013). Playful learning and Montessori education. *American Journal of Play*, *5*(2), 157–186. Retrieved from https://files.eric.ed.gov/fulltext/EJ1003949.pdf

Lillard, A. S., Lerner, M. D., Hopkins, E. J., Dore, R. A., Smith, E. D., & Palmquist, C. M. (2013). The impact of pretend play on children's development: A review of the evidence. *Psychological Bulletin*, *139*(1), 1–34.

Lindell, M. K., House, D. H., Gestring, J., & Wu, H. C. (2018). A tutorial on DynaSearch: A Web-based system for collecting process-tracing data in dynamic decision tasks. *Behavior Research Methods*, 1–15.

Loh, C. S. (2012). Information trails: In-process assessment of game-based learning. In D. Ifenthaler, D. Eseryel, & X. Ge (Eds.), *Assessment in game-based learning* (pp. 123–144). New York: Springer.

Lord, F. M. (1968). Some test theory for tailored testing. *ETS Research Bulletin Series, 1968*(2), i–62.

Malone, T. W., & Lepper, M. R. (1987). Making learning fun: A taxonomy of intrinsic motivations for learning. In R. E. Snow & M. J. Farr (Eds.), *Aptitude, learning, and instruction: Cognitive and affective process analysis* (Vol. 3, pp. 223–253). Hillsdale, NJ: Lawrence Erlbaum Associates. https://doi.org/10.1016/S0037-6337(09)70509-1

Markus, K. A., & Borsboom, D. (2013). *Frontiers in test validity theory: Measurement, causation and meaning*. New York: Psychology Press.

Masters, G. N. (1982). A Rasch model for partial credit scoring. *Psychometrika, 47*(2), 149–174.

McClelland, J. L. (1988). Connectionist models and psychological evidence. *Journal of Memory and Language, 27*(2), 107–123.

McKee, K. L., Rappaport, L. M., Boker, S. M., Moskowitz, D. S., & Neale, M. C. (2018). Adaptive equilibrium regulation: Modeling individual dynamics on multiple timescales. *Structural Equation Modeling: A Multidisciplinary Journal*, 1442224. https://doi.org/10.1080/10705511.2018

McNamara, D. S., & Kendeou, P. (2017). Translating advances in reading comprehension research to educational practice. *International Electronic Journal of Elementary Education, 4*(1), 33–46.

Mislevy, R. J. (2016). How developments in psychology and technology challenge validity argumentation. *Journal of Educational Measurement, 53*(3), 265–292.

Mislevy, R. J., Oranje, A., Bauer, M., von Davier, A. A., Hao, J., Corrigan, S. et al. (2014). *Psychometric considerations in game-based assessment*. New York: Institute of Play.

Mitchell, M. (2009). *Complexity: A guided tour*. New York: Oxford University Press.

Molenaar, P. C. M. (2004). A manifesto on psychology as idiographic science: Bringing the person back into scientific psychology, this time forever. *Measurement, 2*(4), 201–218.

Molenaar, P. C. M. (2014). Dynamic models of biological pattern formation have surprising implications for understanding the epigenetics of development. *Research in Human Development, 11*(1), 50–62.

National Research Council. (2012). *A framework for K-12 science education: Practices, crosscutting concepts, and core ideas*. Washington, DC: National Academies Press.

Ocone, A., Millar, A. J., & Sanguinetti, G. (2013). Hybrid regulatory models: A statistically tractable approach to model regulatory network dynamics. *Bioinformatics, 29*(7), 910–916.

Papert, S., & Harel, I. (1991). *Constructionism*. Norwood, NH: Ablex Publishing Corporation.

Quellmalz, E. S., Davenport, J. L., Timms, M. J., DeBoer, G. E., Jordan, K. A. et al. (2013). Next-generation environments for assessing and promoting complex science learning. *Journal of Educational Psychology, 105*(4), 1100–1114.

Raudenbush, S. W., & Bryk, A. S. (2002). *Hierarchical linear models: Applications and data analysis methods*. Thousand Oaks, CA: Sage.

Reckase, M. D. (2009). *Multidimensional item response theory*. New York: Springer.

Resnick, L. B., & Resnick, D. P. (1992). Assessing the thinking curriculum: New tools for educational reform. In B. Gifford & M. O'Connor (Eds.), *Changing assessments* (pp. 37–75). New York: Springer.

Resnick, M. (2006). Computer as paintbrush: Technology, play, and the creative society. In D. Singer, R. Golikoff, & K. Hirsh-Pasek (Eds.), *Play = learning: How play motivates and enhances children's cognitive and social-emotional growth* (pp. 192–208). New York: Oxford University Press.

Resnick, M. (2008). Sowing the seeds for a more creative society. *Learning & Leading with Technology, 35*(4), 18–22.

Rogoff, B. (1990). *Apprenticeship in thinking: Cognitive development in social context.* New York: Oxford University Press.

Roth, W. M. (1998). Situated cognition and assessment of competence in science. *Evaluation and Program Planning, 21*(2), 155–169.

Russ, S. W. (2003). Play and creativity: Developmental issues. *Scandinavian Journal of Educational Research, 47*(3), 291–303.

Schilling, S. G. (2007). The role of psychometric modeling in test validation: An application of multidimensional Item Response Theory. *Measurement: Interdisciplinary Research and Perspectives, 5*(2–3), 93–106.

Schwartz, D. L., & Arena, D. (2013). *Measuring what matters most: Choice-based assessments for the digital age.* Cambridge, MA: MIT Press.

Shaffer, D. W. (2006). *How computer games help children learn.* New York: Palgrave Macmillan.

Shaffer, D. W., & Gee, J. (2012). The right kind of gate: Computer games and the future of assessment. In M. Mayrath, J. Clarke-Midura, & D. Robinson (Eds.), *Technology based assessment for 21st century skills: Theoretical and practical implications from modern research.* New York: Springer-Verlag.

Shaw, R. E., & Turvey, M. T. (1999). Ecological foundations of cognition: II. Degrees of freedom and conserved quantities in animal-environment system. *Journal of Consciousness Studies, 6*(11–12), 111–123.

Shute, V., & Ke, F. (2012). Games, learning, and assessment. In D. Ifenthaler, D. Eseryel, & X. Ge (Eds.), *Assessment in game-based learning: Foundations, innovations and perspectives* (pp. 43–58). New York: Springer.

Siddiq, F., Gochyyev, P., & Wilson, M. (2017). Learning in digital networks—ICT literacy: A novel assessment of students' 21st century skills. *Computers & Education, 109*, 11–37.

Skinner, B. F. (1988). Preface to the behavior of organism. *Journal of the Experimental Analysis of Behavior, 50*(2), 355–358.

Squire, K. (2006). From content to context: Videogames as designed experience. *Educational Researcher, 35*(8), 19–29.

Squire, K. (2008). Video games and education: Designing learning systems for an interactive age. *Educational Technology*, 17–26.

Steinkuehler, C., & Squire, K. (2014). Videogames and learning. In K. Sawyer (Ed.), *Cambridge handbook of the learning sciences* (2nd ed.). New York: Cambridge University Press.

Stephen, D. G., Boncoddo, R. A., Magnuson, J. S., & Dixon, J. A. (2009). The dynamics of insight: Mathematical discovery as a phase transition. *Memory & Cognition, 37*(8), 1132–1149.

Templin, J., & Bradshaw, L. (2014). Hierarchical diagnostic classification models: A family of models for estimating and testing attribute hierarchies. *Psychometrika, 79*(2), 317–339.

Tsai, F.-H. (2018). The development and evaluation of a computer-simulated science inquiry environment using gamified elements. *Journal of Educational Computing Research, 56*(1), 3–22.

Van der Linden, W. J., & Glas, C. A. (Eds.). (2000). *Computerized adaptive testing: Theory and practice.* Dordrecht: Kluwer Academic.

Vera, A. H., & Simon, H. A. (1993). Situated action: A symbolic interpretation. *Cognitive Science, 17*(1), 7–48.

Vogel, J. J., Vogel, D. S., Cannon-Bowers, J., Bowers, C. A., Muse, K., & Wright, M. (2006). Computer gaming and interactive simulations for learning: A meta-analysis. *Journal of Educational Computing Research, 34*, 229–243.

von Davier, A. A. (2017). Computational psychometrics in support of collaborative educational assessments. *Journal of Educational Measurement, 54*(1), 3–11.

von Davier, A. A., & Halpin, P. F. (2013). Collaborative problem solving and the assessment of cognitive skills: Psychometric considerations. *ETS Research Report Series, 2013*(2), i–36.

Vygotsky, L. S. (1966). Play and its role in the mental development of the child. *Soviet Psychology, 12*(6), 62–76. Retrieved from https://files.eric.ed.gov/fulltext/EJ1138861.pdf

Vygotsky, L. S. (1978). *Mind in society: The development of higher psychological processes*. Cambridge, MA: Harvard University Press. Retrieved from http://home.fau.edu/musgrove/web/vygotsky1978.pdf

Wiggins, G. P. (1993). *Assessing student performance: Exploring the purpose and limits of testing*. San Francisco, CA: Jossey-Bass.

Wilkerson, M. H., Shareff, R., Laina, V., & Gravel, B. (2018). Epistemic gameplay and discovery in computational model-based inquiry activities. *Instructional Science, 46*(1), 35–60. Retrieved from www.ocf.berkeley.edu/~mwilkers/wp-content/uploads/2017/10/WilkSharLainGrav2017_PP.pdf

Wilson, K. A., Bedwell, W. L., Lazzara, E. H., Salas, E., Burke, C. S., Estock, J. L., . . . Conkey, C. (2009). Relationships between game attributes and learning outcomes: Review and research proposals. *Simulation & Gaming, 40*, 217–266. https://doi.org/10.1177/1046878108321866

Wilson, M., Gochyyev, P., & Scalise, K. (2017). Modeling data from collaborative assessments: Learning in digital interactive social networks. *Journal of Educational Measurement, 54*(1), 85–102.

Wouters, P., van Nimwegen, C., van Oostendorp, H., & van der Spek, E. D. (2013). A meta-analysis of the cognitive and motivational effects of serious games. *Journal of Educational Psychology, 105*, 249–265.

Young, M. F. (1993). Instructional design for situated learning. *Educational Technology Research and Development, 41*(1), 43–58.

Young, M. F. (1995). Assessment of situated learning using computer environments. *Journal of Science Education and Technology, 4*(1), 89–96.

Young, M. F. (2004). *An ecological description of video games in education*. Paper presented at the International Conference on Education and Information Systems Technologies and Applications (EISTA), Orlando, FL, July 2004.

Young, M. F., & Barab, S. A. (1999). Perception of the raison d'etre in anchored instruction: An ecological psychology perspective. *Journal of Educational Computing Research, 20*(2), 119–141.

Young, M. F., DePalma, A., & Garrett, S. (2002). An ecological psychology perspective on situations, interactions, process and affordances. *Instructional Science, 30*, 47–63. Retrieved from http://web.uconn.edu/myoung/InstSciY,D,&G2002.pdf

Young, M. F., Kulikowich, J. M., & Barab, S. A. (1997). The unit of analysis for situated assessment. *Instructional Science, 25*(2), 133–150.

Young, M. F., & Slota, S. (2017). *Exploding the castle: How video games and game mechanics can shape the future of education*. Charlotte, NC: Information Age Publishing.

Young, M. F., Slota, S., Cutter, A. B., Jalette, G., Mullin, G., Lai, B. et al. (2012). Our princess is in another castle a review of trends in serious gaming for education. *Review of Educational Research*, *82*, 61–89. Retrieved from http://journals.sagepub.com/doi/10.3102/0034654312436980

Zheng, D., Newgarden, K., & Young, M. F. (2012). Multimodal analysis of language learning in World of Warcraft play: Languaging as values realizing. *ReCALL*, *24*, 339–360. Retrieved from www.academia.edu/1979055/Zheng_D._Newgarden_K._and_Young_M.F._2012_._Multimodal_analysis_of_language_learning_in_World_of_Warcraft_play_Languaging_as_values_realizing._ReCALL_24_339-360

4 Improving Situation Awareness in Social Unrest Using Twitter

A Methodological Approach

Peter K. Forster and Samantha Weirman

CONTENTS

INTRODUCTION

In today's highly interconnected society, social media's ability to motivate groups and individuals to action, for both good and bad, is evident. Russia's interference in the 2016 election both in terms of longevity of activity and goals offers a poignant example. Evidence exists that Russia may have penetrated various systems including the Clinton campaign and the DNC as early as 2012 (Zegart & Morrel, 2019). While intrusion of the systems in 2016 was detected, its potential impact on the election was not publicly acknowledged until a month before the election (Zegart & Morrel, 2019). Russia's "weaponization" of social media reflects the technical and social coalescence known as computational propaganda, or the ability to manipulate public opinion (Bolsover & Howard, 2017). Propagandists target emotions, seeking to influence perspective and behavior and often looking to inspire mobilization among the recipients and consumers of their content. The Russian case demonstrates the analysis of content, device, and network impact on individuals and groups is in its infancy.

Burgeoning open source intelligence (OSINT), available through social media and interconnectivity, demands a re-evaluation of how information is gathered and intelligence created. This chapter embraces the notion that networks are constantly changing based upon events and the framing and dissemination of information used to interpret those events. It thus contributes to a better understanding of an evolving intelligence environment. Within this study, the authors present examples to demonstrate how synthesized Twitter content reflects real-world events. Specifically, this chapter explores the relationship between patterns of Twitter communications and associated offline (i.e. real-world) events during civil unrest. It demonstrates the alignment of popular words and phrases with real-world events, offering a proof of concept for real-time narrative construction through n-gram and noun phrase extraction. It postulates that tracking patterns in online social media data can improve offline situation awareness and decision making by providing insights into what people are talking about, how they are organizing and contextualizing their messages, and how people feel about the topics and themes being discussed. In essence, synthesis of individual social media messages may produce a story from which indicators and warnings of group-level behavior can be derived.

This chapter leverages a Twitter data set related to the 2017 Unite the Right (UTR) rally in Charlottesville, Virginia, in order to better understand the relationship between Twitter communications and mobilization culminating in threats and frictions between potentially violent groups. First, a linear timeline of relevant events and activities occurring between August 8 to August 15, 2017 is presented and compared to the corpus of tweets spanning that same period. To establish a context, the chapter offers an abbreviated overview of Twitter's role in creating influence and motivating collective action with a particular focus on the methodologies used. Building upon this foundation, it lays out an event-driven methodology which juxtaposes physical mobilization (offline activities) with Twitter data (online conversations). To parse the data further, the timeline is broken down into pre-event, event, and post-event time frames. Within these periods, this study explores what stories are being crafted by analyzing tweet characterizations and n-grams, or word sequences. Specifically, it examines the following questions.

- What are the Twitter communication patterns in the pre-event environment?
- What are the Twitter communication patterns at times of heightened threats and frictions?
- What are the Twitter communication patterns in a post-event environment?

This analysis has potential to improve situation awareness. First, it is demonstrated that even without prior knowledge of an event, periodic n-gram and noun phrase extraction from Twitter conversations related to that event reveals a coherent narrative. From the UTR Twitter data set, the "story" crafted through this process aligns with a linear timeline of real-world events, developed from official reporting and coverage of the event. Additionally, Twitter data also adds a level of intelligence about emotion and sentiment that can positively improve decision making. The fusion of information derived from social media and through traditional methods of data collection creates an opportunity to identify indicators and warnings of potential violence. Finally, by separating online communications from their creators (or original publishers), the focus of awareness is redirected from individual players to the underlying environment and wider network of actors who may be united by these messages. This reprioritization acknowledges the reality that messages can and often do become bigger than one person; they essentially take on a life of their own.

CREATING SITUATION AWARENESS FROM TWITTER

At a fundamental level, situation awareness (SA) is deriving what is important from what is going on around you (Endsley & Garland, 2000). It is about perceiving the cues or important pieces of information contained within a data-rich environment in order to integrate and comprehend a current state and establish a foundation for understanding a future one (Endsley & Garland, 2000). In essence, SA reflects the collection, prioritization, and analysis phases of the intelligence process. By definition, SA is a snapshot in time; however, to be truly effective, the SA processes should be integrated across time to understand how the environment is changing (Endsley & Garland, 2000). Today's interconnectivity demands that SA consider physical, as well as virtual, cues. Within this context, analyzing Twitter activity can provide added dimensions about how people are organizing and what is important to them.

Communications and networking over virtual channels, such as social media, should not be considered independent of what happens on the ground (Carter, Maher, & Neumann, 2014). In fact, analysis of online content provides valuable insight into motivations behind an observable behavior and the direction—growing or waning—of the sentiment influencing past and present real-world activities. Understanding motivations and fervor of sentiments, particularly across time, offers a foundation for predicting future real-world events. Consequently, consideration of this interconnectivity has affected contemporary understanding of extremist groups, actions, and schools of thought. Experts in the field widely recognize the general, online presence of extremists, as well as the importance these entities place on virtual communication channels to mobilize people for physical action (Ashcroft, Fisher, Kaati, Omer, & Prucha, 2016; Frampton, Fisher, & Prucha, 2017; Vidino & Hughes, 2015).

Although trailing jihadists in effective use of online media, right-wing extremist connectivity is growing and provides a stark example of the real-virtual world connections. The Identitarian Movement, seeking to establish and mobilize a global network, uses #identitarian to connect like-minded nativists in Scandinavia, Italy, Germany, Australia, and the United States (The Economist, 2018). Trending topics offer insight into what is important to the group and may yield information on future action. Analyzing virtual comments made by perpetrators of political or religious violence is one method that can be used to improve situation awareness. For example, the gunman in the 2019 Christchurch mosque shootings referenced the writings of Dylan Roof and particularly "Knight Justiciar" (Andres) Breivik as inspiring his actions before livestreaming his attack (Taylor 2019). Identifying others who cite and support Breivik or Dylan may provide clues on future attacks. In another example, the 2019 Earle Cabell Federal Courthouse shooter made multiple comments on Facebook and posted a YouTube video depicting him with a rifle and announcing "a storm is coming" in the days preceding the attack (CBSDFW, 2019). Although he did not offer a clear date, time, or target of the attack in these messages, his virtual behavior and postings indicated someone preparing for action.

While a foundational principle of offline and online interconnectedness serves this chapter well, a gap exists in understanding the relationship between social media usage and physical activity, specifically in how one influences the other.

Althoff, Jindal, and Leskovec (2017) studied 6 million users over five years, focusing on individual interaction with social media and changes in offline activity. In this study, the authors found behavioral change occurred over a long time period, effectively limiting value in real-time situation awareness improvement. Perhaps more closely aligned to this chapter, the Bui (2016) case study examined Internet penetration in the population and established that social media enlarged the political space. However, the focus of the study was on technologies, rather than content. The study presented in this chapter deviates from these approaches by leveraging Twitter data to examine content and its relationship to what is happening in the real world across a defined time period.

Additionally, this chapter discusses the challenge of establishing what determines influence in the online space. While identifying Twitter users who are radicalized is relatively easy, deciphering who is important and worth watching is more difficult (Berger & Strathearn, 2013). In studying ISIS supporters, Berger and Morgan (2015) began with "seed accounts" of known influential people and then used social network analysis (i.e., connectivity among individuals, paths to expedite message transmission, and centrality or importance of certain individuals) and daily monitoring of Twitter to ascertain supporters' validity and eliminate "noise" in the system. The study used an individual's network position as an indicator of importance. Carter et al. (2014) also connected influence to network position by examining a node's in-degree centrality or who followed, retweeted, or mentioned a specific node, typically an individual. This chapter takes a different approach and "reverse engineers" the process by first identifying popular messages and then working backwards to see where and how they originated. Furthermore, it analyzes the messages in the context of real-world activities. This method considers both how and why a message or conversation changes over the course of an event and uses this information to better assess their importance.

This chapter also addresses a need for empirical research with sufficient sample size. Certainly, Berger and Carter's approaches have credence, but neither study was done at scale. Berger and Strathearn's "seed account" study analyzes 342,807 tweets and Carter, Maher, and Neumann only use 14,509 tweets in their work. The Charlottesville case presented in this chapter examines over 5 million tweets collected from August 8 to August 15, 2017. Notably, extracting analytically valuable information from such a large data set requires a less manually intensive process, thus the study uses a combined qualitative-quantitative approach. Ashcroft et al. (2016) applied machine learning and sentiment classification to classify a large number of tweets as supporting ISIS, but they readily admit that their data parsing does not determine radical content or purpose of the message. Thus, while their study helps to sort through Twitter noise—a challenge identified by Berger—distinguishing radical content requires refined and increased SA, particularly at the tactical level. This chapter introduces an approach to analyzing influence in social media networks that acknowledges that new influencers may emerge during an event, with an additional possibility of a network influencer being a message, rather than a person. Giesea (2017) considers this notion within the context of memes. His basic premise that messages have social effects in shaping ideas is particularly relevant for studies on extremism, where a message often needs to outlive its creator (Weirman & Alexander, 2018).

A growing volume of literature exists that examines the online extremist phenomenon; however, extremist use of the online environment is not a recent phenomenon. Jihadist extremists have been leveraging the online environment for nearly 20 years. While many are aware of the so-called Islamic States' use of social media to recruit followers and exploit events, few understand either the history of the cyber-jihad or its longevity. Long before #ISIS appeared in June 2013, Osama bin Laden was using websites and discussion forums to spread Al-Qaeda's message and exert control over his far-flung organization (Watts, 2018 p. 43). September 11 hijackers Mohammed Atta and Ramzi Binalshibh corresponded via email in August 2001 (The National Commission on Terrorist Attacks Upon the United States, 2004, p. 248). By 2004, "Irahabi 007" (aka Younis Tsouli) was hacking credit card accounts and online photos posted by US troops in Iraq to share their locations, teaching other jihadist how to hack, setting up and monitoring chat rooms, and creating websites with downloads on how to make explosives for Al-Qaeda in Iraq to exploit their successes and maintain connectivity with Al-Qaeda's leadership. Al-Shabaab, the Somali-based jihadi group, embraced Twitter as early as 2007 to connect with the Somali diaspora and to solicit funds (Watts, 2018, p. 51). In 2013, Al-Shabaab livestreamed its attack on the Westgate Mall in Nairobi, Kenya using YouTube (Watts, 2018, p. 54).

Notwithstanding the number of studies impacting this environment, the existing challenges of pinpointing the relationship between events and Twitter messaging, determining who or what is influential, and further developing the concept of messages in the role of network influencers still exist. This chapter seeks to offer a methodology that mitigates these challenges.

BACKGROUND: CHARLOTTESVILLE, AUGUST 2017

In May 2017, Jason Kessler, a Charlottesville resident and member of the alt-right group Proud Boys applied for a permit to conduct a demonstration in Emancipation Park Charlottesville, Virginia to protest the renaming of the park, previously known as the Market Street or Lee Park, and the planned removal of a statue of Robert E. Lee (Jason Kessler v. City of Charlottesville and Maurice Jones, 2017). Upon receiving a permit, the date for his "Unite the Right" demonstration was set for August 12. As the UTR rally was meant to attract a coalition of right-wing groups, it became "increasingly Nazified" and thus more radical (Lind, 2014). The estimated number of attendees, potential presence of firearms among the groups that would be present, and possibility of counter-protesters coming to Charlottesville raised public safety concerns. The Southern Poverty Law Center characterized the demonstration as the largest "hate gathering" in decades in the United States, and Charlottesville Police Department (CPD) intelligence estimated more than 1,000 participants would attend, even drawing people from as far as Texas and California (Jason Kessler v. City of Charlottesville and Maurice Jones, 2017). Kessler admitted the event might significantly exceed the original 400 participants, as he had produced a number of podcasts and Twitter promotions on websites, such as 4chan, to promote the event (Jason Kessler v. City of Charlottesville and Maurice Jones, 2017). Additionally, there was increasing concern about attending groups such as the Fraternal Order of Alt-Knights, who intended to carry firearms, and 150 security personnel from the

Alt-Knights and The American Guard (Jason Kessler v. City of Charlottesville and Maurice Jones, 2017). Kessler also fed the fear of violence by referring to the 2017 clashes between the left-wing extremists, called Antifascists (Antifa), and right-wing groups at University of California, Berkeley (Jason Kessler v. City of Charlottesville and Maurice Jones, 2017). Faced with converging unpalatable dynamics, the City of Charlottesville rescinded the original permit and made a public announcement that it was seeking to move the "rally" on August 7, 2017.

Between August 8 and 10, friction grew between Kessler and city leaders. After an August 8 meeting between Kessler and the chief of police, Kessler tweeted, "We go either way" announcing that the rally would take place in Emancipation Park regardless of city action (The Rutherford Institute, 2017). At 19:00 on August 11, the University of Virginia police (UPD) discovered social media posts discussing marching through the University of Virginia campus (Hunton & Williams LLP, 2017). On August 11, Kessler announced a lawsuit in US District Court to "reinstate" his permit via Twitter (Jason Kessler v. City of Charlottesville and Maurice Jones, 2017). During the same time frame, additional frictions emerged. Concerned with the rally's growing neo-Nazi direction, the Congregation of Beth Israel requested security protection from CPD (Hunton & Williams LLP, 2017). Having, in essence, drawn a "line in the sand," it seemed inevitable that the rally would take place at Emancipation Park and the city began preparations for the event.

Late on the evening of August 11, members of Unite the Right started a torch-lit march through the University of Virginia (UVA) arriving at the rotunda in the center of campus. Although some pushing and shoving occurred, the UPD, with assistance from CPD, was able to disperse the crowd for the evening. However, the respite was short-lived. At 9:00 on the morning of August 12, Jack Pierce, the head of Unite the Right security, refused a CPD escort into Emancipation Park, as Eli Mosely simultaneously led the neo-Nazi group Identity Evropa in Emancipation Park (Hunton & Williams LLP, 2017). By 10:00, fights began to break out on the streets surrounding the park. Some UTR members were arrested for throwing stones, while others clashed with Antifa, who were fighting to keep them out of Emancipation Park (Hunton & Williams LLP, 2017). As CPD, the Virginia State Police (VSP), and the Virginia National Guard (VaNG) tried to disperse the groups, chaos began to spread, and violence escalated. One counter-protester, Corey Long, fired an improvised flamethrower at members of UTR. In response, Richard Preston, a UTR member, fired a pistol at Long's feet (Hunton & Williams LLP, 2017). Soon after, DeAndre Harris was assaulted by UTR participants in the parking garage. Around 13:40, violence reached its apex as James Fields drove a car into a crowd of people at 4th and Water Streets, killing counter-protestor Heather Heyer and injuring 28 others (Hunton & Williams LLP, 2017).

Fields' ramming attack changed the complexity of the event. Shortly after the attack, Fields was arrested while the VSP and VaNG moved to secure the crime scene and evacuate injured victims. In a last call for more violence, Twitter posts around 15:00 urged firebombing of the Beth Israel Congregation (Hunton & Williams LLP, 2017). Increased police presence and the subsequent declaration of a curfew dispersed the crowd as the afternoon progressed and police demobilized by 17:00 (Hunton & Williams LLP, 2017). At 14:00 on August 13, Kessler held a press conference to address the events of the prior day but ultimately needed to be escorted to CPD headquarters for his own safety. At this point, the physical event concluded, but a virtual post-event period emerged as Twitter continued to be a forum for engagement, threats, and sympathy.

METHODOLOGY

As previously indicated, this chapter studies online influence differently. Rather than looking at connections from seed accounts, as Berger and Strathearn did, this work is more reflective of Frampton, Fisher, and Prucha's concept of "swarmcast," an interconnected network that displays agility, speed, and resilience allowing itself to be reconfigured (Frampton et al., 2017). This approach embraces the notion that networks are constantly evolving based upon external stimuli but does not go into depth to understand which stimuli may be more influential. This study seeks to remedy this shortcoming. Here, the researchers examined linkages between physical real-word events, or what is happening offline, and online communication, or what is occurring on Twitter. This research leverages a time-bound Twitter data set related to the 2017 Unite the Right rally and responses. It includes over 5 million public tweets by more than 1.5 million unique users posted between the dates of August 8, 2017 through August 15, 2017.

With a framework established, the study migrated from a theoretical analysis of what might go on to a data-driven one. Two distinct data sets collected from Twitter's API allowed researchers to derive a time-bound corpus of tweets. Justin Littman originally curated the first data set using the terms #charlottesville, #standwithcharlottesville, #defendcville, #heatherheyer, and #unitycville (Littman, 2018). Ed Summers curated the second data set with #unitetheright (Summers, 2017b). The Hydrator application was used to expand each data set from the list of the tweet IDs provided (Summers, 2017a). As is evident from the hashtags, these two data sets broaden the research opportunities by introducing differing perspectives. Researchers combined the data sets, removing duplicate tweets, to get a more comprehensive picture of what was occurring in Charlottesville.

To test the validity of using social media data to improve SA of real-world events, this study leveraged the Hunton and Williams LLP (2017) After Action Report (AAR) on Charlottesville to compare the official timeline of events against the Twitter data set. The researchers first established date/time blocks using hourly "ceiling" values (e.g., 8/9/2017 20:00 covered all events and Twitter traffic from 19:00:01 to 20:00:00 on August 9). The time frame of interest—8/8/2017 1:00 through 8/16/2017 00:00—was then divided into pre-event (8/8/2017 1:00 to 8/11/2017 20:00), event (8/11/2017 21:00 to 8/12/2017 20:00), and post-event (8/12/2017 21:00 to 8/16/2017 00:00) periods, to allow for a more refined categorization and analysis of data. These divisions were informed by the AAR. The pre-event period is bounded by the time when Jason Kessler informed CPD that he intended to hold a rally at Emancipation Park regardless of the city's action on August 11 to when CPD is told by Kessler's representative that the march will begin at Nameless Field (Hunton & Williams LLP, 2017). The event period starts 45 minutes before the march on the night of August 11 and ends one hour after CPD's demobilization on August 12. The post-event period commences immediately after and continues until 08:00:00 on August 15. Analysis of Twitter data was performed on the whole data set in its entirety, for the three defined time periods (pre-event, event, post-event), and by hourly interval.

After creation of the time buckets, researchers considered the number of tweets and message characterization over time, determining whether the message was original, retweeted, quoted, or "shared" using a website's tweet button. As will be discussed later, this evaluation provides insight as to how people use Twitter to respond to an event. It also raises the question regarding motivation for sharing a particular message. Ultimately, this question deserves greater analysis that is beyond the scope of this chapter. Next, word and phrase extraction were used to identify the most commonly used phrases in the network. Popular content provides a shortcut for acquiring a general sense of what is happening on the ground without consulting external sources. Finally, researchers looked at sentiment of the tweets in the network, using VADER's composite sentiment score algorithm (Hutto & Gilbert, 2014). Results were visualized with Tableau (Tableau Desktop, 2003).

RESULTS AND DISCUSSION

Analysis of Twitter activity and the concepts or phrases that gained traction provides some insight into how the network is engaging in conversation, what people are discussing, and how they feel about it. The resulting data fusion approach improves situation awareness by illuminating level of activity and mobilization associated with network behavior, conversation, and popular messages on the Twitter platform.

To start, researchers looked at Twitter activity in the days before, during, and after the event. As anticipated, Figure 4.1 indicates that the number of related

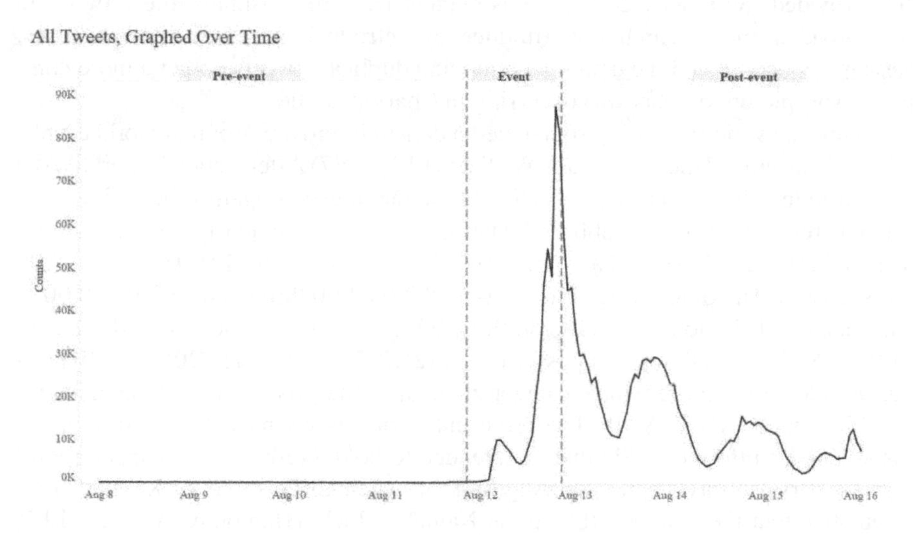

FIGURE 4.1 Hourly count of tweets related to the 2017 Unite the Right event in Charlottesville, Virginia, from August 8, 2017 through August 15, 2017.

tweets was limited during the pre-event stage, notwithstanding extensive planning by Charlottesville authorities and negotiations with the UTR rally organizer, Jason Kessler. Activity accelerates as the rally starts and confrontation between the UTR and counter-protestors grows. Activity initially peaks with over 50,000 tweets during Fields' vehicle ramming attack that killed Heather Heyer and injured 28 others. Twitter activity briefly drops, perhaps as people pause to comprehend what has occurred, before skyrocketing to its highest number of nearly 90,000 tweets near the end of the event time period (i.e., 20:00 on August 12). The post-event time frame reflects a consistent downward trend on Twitter from its high number of approximately 55,000 tweets in the evening of August 12. The spikes are likely responses to various White House comments, including Jeff Session's calling Fields' attack "domestic terrorism" (Astor, Caron, & Victor, 2017).

Figure 4.2 classifies the total number of tweets into four message categories that indicate how people use Twitter to spread messages. "Original" messages are created by the person who posts them on Twitter. These individuals may or may not actually be at an event. "Retweets" are a re-posting of a message that has already been shared, by either the retweeting account or another user. Retweets enable further dissemination of a message to new audiences, thus expanding the message's potential impact. "Quoted tweets" are a curation process in which the poster quotes someone else's tweet within their message but adds their perspectives or commentary. Finally, messages may be shared with Twitter buttons displayed on a website or other platform. The share functionality makes it easy for Twitter users to post a message to their feed and makes it possible for otherwise unconnected Twitter users to tweet the same message. This process, again, increases potential amplification of a single, unified message.

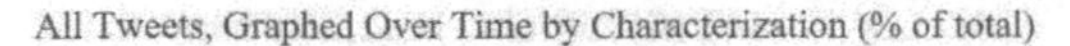

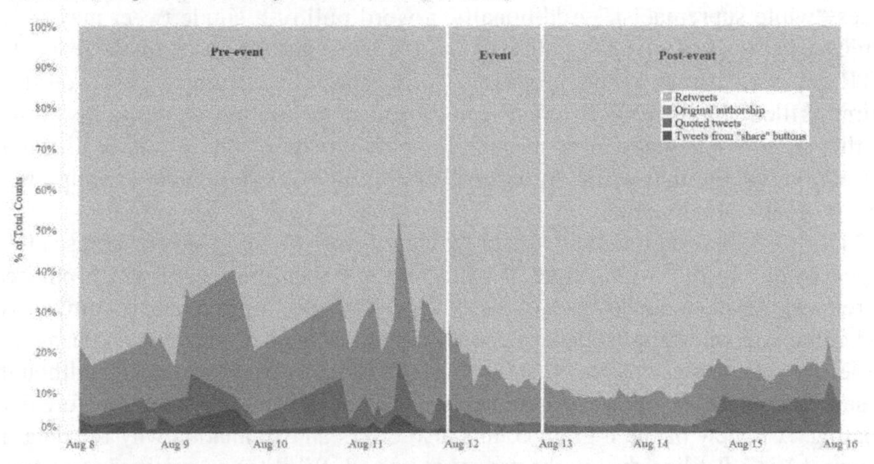

FIGURE 4.2 Hourly counts of tweet characterization, plotted as a percentage of overall Twitter activity, related to the 2017 Unite the Right event in Charlottesville, Virginia, from August 8, 2017 through August 15, 2017.

The value of this data is in helping to recognize the importance of the message itself, as well as the level of engagement among Twitter users. Examination of retweets, quotes, and shares helps determine which messages resonate with a community, understand the evolution of narratives, assess when a message has assumed a life of its own beyond the original creator(s) of the message, and identify influencers in the network. This tweet behavior provides insight regarding shared situation awareness, or the degree to which individual entities within the network possess the same knowledge and perception of events. Furthermore, the method of posting tweets reflects network engagement. Newly authored and quoted messages indicate higher levels of engagement with the message, whereas retweets and shared content require little effort on the part of the posting user.

As is common with most data sets, the Charlottesville's data is dominated by retweets. Retweet activity points to what messages are sticking within the user community. Notwithstanding, the pre-event period reflects more original activity, as defined by original, quoted, and shared tweets, and thus represents an engaged community. People are sharing ideas and expressing opinions. Although somewhat contradictory to what was reflected in Figure 4.1, which showed little Twitter activity, the overall number of tweets might explain it. Over time, original messaging dips a bit but remains relatively consistent. Conversely, quoted and shared messages decline significantly over time, particularly between midday on August 12 and the end of August 14.

Researchers were also interested in how content could provide clues about what was happening "on the ground." This study leveraged noun phrase, single word frequency, and n-gram (specifically, bi- and trigram) extraction. N-grams refer to a sequence of words, such as "unite the right." The decision to assess content in word groupings of one (single word frequency), two (bigrams), and three (trigrams) was made to account for word coupling, order, and arrangement. Single-word frequency extractions will capture "white" and "supremacist," but multi-word provides "white supremacist." Additionally, a word pull of a single tweet may yield three keywords but give no indication of the ordering. In an example from the UTR data set, "trump, killed, protestor," "protestor, killed, trump," and "protestor, trump, killed" imply very different scenarios, despite containing the same words. In this case, it was made clear through the comparison of the results of different extraction lengths that while "protestor" and "killed" were related, "trump" was part of a separate thought.

Figure 4.3 dissects the Twitter data by identifying the top words or phrases that appeared during the overall time of collection and then presenting which phrases were most prevalent during each of the time periods. The n-gram data is cumulative and collapsed, when applicable (e.g., combining hate and hatred counts). During the pre-event period, n-grams resulting from UTR hashtags were predominant, although in small numbers. The pre-event momentum carried over into the march before dissipating. A review of the n-gram counts also offers an explanation why the data in Figure 4.1 is a flat line during the pre-event period. While conversations were happening, they were so few in comparison to event and post-event periods that it was impossible to display the activity within a readable graphic.

Most Popular N-Grams Overall (Top 10)

	Count	% of total tweets overall with n-gram
white	937,499	18.65%
hate OR hatred	571,373	11.37%
supremacists OR supremacy	460,801	9.17%
white supremacists OR white supremacy	438,943	8.73%
trump	407,890	8.11%
nazi	319,507	6.36%
violence	292,746	5.82%
people	224,669	4.47%
rally	212,465	4.23%
racism	212,175	4.22%

Pre-event

	Count	*
rally	1,299	22.47%
right	945	16.35%
park	550	9.51%
charlottesville	536	9.27%
white	459	7.94%
lee park	435	7.52%
city	429	7.42%
permit	406	7.02%
neonazi	397	6.87%
altright	349	6.04%

Event

	Count	*
white	216,829	19.47%
nazi	185,928	16.70%
hate OR hatred	140,035	12.58%
right	113,335	10.18%
trump	112,214	10.08%
rally	87,741	7.88%
violence	70,384	6.32%
display	67,504	6.06%
people	67,491	6.06%
police	55,239	4.96%

Post-event

	Count	*
white	702,645	17.98%
nazi	492,482	12.60%
hate OR hatred	436,198	11.16%
trump	378,624	9.69%
supremacists **	362,979	9.29%
white supremacists **	354,027	9.06%
side	336,224	8.60%
nazis	267,362	6.84%
violence	219,232	5.61%
racism	166,821	4.27%

** OR supremacy

*** % of tweets in each period with n-gram**

FIGURE 4.3 Most popular n-grams in tweets related to the 2017 Unite the Right event in Charlottesville, Virginia, from August 8, 2017 through August 15, 2017, cumulative and during each period.

In the pre-event environment, three salient points emerge from the comparison of the Twitter communication patterns and real-world events. First, while the UTR's messaging is dominant in the Twitter-sphere, it has limited resonance and no connection to Kessler. Second, while engagement was limited, the amount of original content, defined as original or quoted posts, was relatively high. This level of originality implies single messages are neither sticking nor mobilizing. Thus, there appears to be limited concern of confrontation. Third, despite some incendiary events, such as Kessler's tweet about rallying in Emancipation Park regardless of city action, there is limited reference to threatened violence and possibility of friction appears low. However, by the end of the day on August 9, #defendcville has joined #unitetheright and #charlottesville as the third most popular hashtag in the Twitter-sphere. From an analyst's perspective, this appears to indicate the emergence of an oppositional group that should require some exploration. Who makes up this group? Are known violent counter-protestors in the area?

The most popular n-grams were "rally," "right," "park," and "lee park," with "rally" accounting for 22% of n-grams over this time period. The terminology used is relatively void of emotion and instead reflects organization, which indicates that UTR members were more active than counter-protestors were during this period. Although not unexpected, Twitter was relatively quiet during this time despite the planning that was taking place. Many of the conversations and meetings during this time period were either not public or lacked wide public appeal. Still, several potentially confrontational activities including tweets by Kessler, other social media postings about the rally, requests for protection from the Congregation of Beth Israel, and growing police and VaNG presence on August 11 occurred. None of these activities gained significant traction on Twitter, supporting conclusions that while the number of rally participants far exceeded expectations, there was little coordination or mobilization taking place in the virtual space. Furthermore, the lack of response to Kessler's tweet demonstrates that despite Kessler's formal leadership role, his activities had minimal impact at this point, even when amplified by social media. From a situation awareness perspective, this finding indicates that the event itself was much bigger than the personalities behind it. Finally, the amount of original content being posted may suggest people are trying to understand what is about to occur but are not necessarily committed to action. What was not present on Twitter is more interesting than what existed. The lack of emotional appeals by either side—the UTR and the counter-protestors—as well as the lack of influencers was surprising.

As expected, as physical activity increases during the event period, so too does Twitter activity. Terms such as "white," "right," and "rally" continue to appear, but more emotional terms such as "hatred," "violence," and "nazi" are increasingly frequent and compose 35% of the most-mentioned n-grams. The messages indicate that UTR's momentum is dissipating and the counter-protestors are beginning to drive the narrative. At 11:00, law enforcement deploys and an unlawful assembly is declared (Hunton & Williams LLP, 2017). Perhaps law enforcement now thought that tensions were defused. Nevertheless, the volume on Twitter, particularly the predominance of retweets, indicates messages are motivating collective action. This trend corresponds with growing altercations between the UTR and counter-protestors, which climax with Fields' ramming attack about 13:30 and the urging of the firebombing of the Congregation of Beth Israel at 15:00 (Hunton & Williams LLP, 2017). Recognizing this dynamic would be helpful to law enforcement.

The analysis of the growth of retweets and message resonance further validates this theory. The emotional shift that occurred during the event period corresponds with original content being replaced by retweets with specific hashtag or URL identifiers. People coalesced around specific positions and unified messages gained traction. "There is only one side. #charlottesville" and "Even as we protect free speech and assembly, we must condemn hatred, violence and white supremacy. #charlottesville" were tweeted 115,343 and 99,268 times, respectively. At this point, the UTR, perhaps either due to lack of commitment or incentive to hide, ceded the Twittersphere. Furthermore, the Twitter platform facilitated the mobilization of collective engagement and establishment of a brand around which people could coalesce. Commentary in the form of retweets and curation reflected opinions of the network as events unfolded, whereas information dissemination through these same means often promoted action. From a situation awareness perspective, the shift from message generation to message convergence may indicate collective mobilization, which has the potential for confrontation and violence. In the wake of the mass casualty event, as the ramming attack was classified (Hunton & Williams LLP, 2017), and the evacuation of the injured, both the physical and the virtual environments entered the post-event stage.

As law enforcement regained control of the situation and people dispersed, Twitter engagement also declined, but remained more active than in the pre-event stage. The event had created momentum. Most of the Twitter volume was still retweets, although some curation re-emerged. In analyzing the n-grams, the divisiveness of the event remained, and blame was assigned. Nearly 87% of n-grams were emotionally focused, using terms such as "nazi," "nazism," "hate/hatred," and "white supremacy/white supremacist." Post-event data included narrative, reflection, and remembrance. Again, the data over this time period indicates the predominance of the message rather than the people behind it.

N-gram and noun phrase extraction can also be useful for real-time SA because they provide a shortcut for understanding what is happening on the ground. The most popular words and word phrases within a time interval of interest can be used to construct a narrative or piece together a story of events. With just these keywords, it is often possible to get a general sense of what is happening, even without consulting external or official reporting sources. In this way, Twitter can be used to track and understand events as they unfold, straight from the public itself, which is particularly valuable where and when official reporting is lacking or controlled.

Table 4.1 shows a sampling of the most popular n-grams and noun phrases in tweets related to the 2017 Unite the Right event in Charlottesville, Virginia, from August 8, 2017 through August 15, 2017. It illustrates how phrase extraction creates a narrative that lines up with real-world events. From the content of the first two hours on August 8, 2017, one could deduce that an event will occur at Lee Park, despite

TABLE 4.1
Sampling of Most Popular N-Grams and Noun Phrases by Hourly Interval Ceiling

Interval Ceiling	Top n-Grams and Noun Phrases
8/8/2017 1:00	lee park, city council, canceled permit, holding, anyway
8/8/2017 2:00	lee park, city council, canceled permit, holding, anyway
...	
8/8/2017 15:00	demand, allow, rally, emancipation park, lee park
8/8/2017 16:00	demand, allow, rally, emancipation park, city council
...	
8/8/2017 19:00	neo-nazi trolls, showing up armed, kekistan militia
8/8/2017 20:00	alt-right organizers, pictures, adolf hitler, promote, unite the right
8/8/2017 21:00	rally, hate, neo-nazis, kissing cousins, justice, lee park, alt-right organizers
8/8/2017 22:00	rally, new article, need, largest, white nationalist rally, many years
...	
8/11/2017 19:00	rally, charlottesville, unite the right, richard spencer, david duke, planning, tonight
8/11/2017 20:00	rally, unite the right, event, planning, tonight, neo-nazis, breaking, torch lit, charlottesville
8/11/2017 21:00	rally, alt-right, lee park, emancipation park, breaking, judge, grants injunction, unite the right, white supremacists
8/11/2017 22:00	rally, white supremacists, torch march, uva, lee park, neo-nazis, torches, federal judge, reinstates permit, nazis, marching
8/11/2017 23:00	uva, rally, march, alt-right, group, counter-protesters, charlottesville, white nationalists, torchlit, mayor mike signer, statement, beyond disgusted
8/12/2017 0:00	rally, statement, mayor mike signer, beyond disgusted, facebook, torches, march, alt-right, massive brawl, jefferson monument, chemicals, dispersed, thrown, breaks
...	
8/12/2017 15:00	car, protesters, hatred, violence, nazi, white supremacy, condemn hatred
8/12/2017 16:00	hatred, violence, display, racism, car, trump, hate, nazi, white nationalist demonstration, no place, society
...	
8/13/2017 1:00	hatred, trump, white supremacists, today, there is only one side, heather heyer, terrorist attack, white nationalist
8/13/2017 2:00	hatred, trump, white supremacists, today, there is only one side, heather heyer, terrorist attack, white nationalist
8/13/2017 3:00	hatred, nazi, trump, white supremacy, remember, today, heather heyer, killed
...	

a cancelled permit. Over the course of that day, it becomes clear that the event is an alt-right, white nationalist rally. In the official timeline, it is not clear when the permit is rescinded; however, based on this Twitter activity, it is known that this action took place before 1:00 on August 8. The official report notes that at 11:00, the

Charlottesville City Council (CCC) discussed moving the march to an alternative venue with Kessler, but he rejected the proposal and notified them between 15:00 and 16:00 that UTR would march regardless. Notably, online discussions about the permit do not mention the change of venue proposal, and instead frame the situation as an outright cancellation. Additionally, the top n-grams in the Twitter data set over this period reflect an appeal to injustice with emotionally charged words (e.g., demand, allow). This selective reporting and framing of events by Kessler and his network of supporters reveals a calculated attempt to activate and mobilize his base.

At 19:00 on August 11, 2017, "planning" and "tonight" emerge as top words among the Twitter network, and by 22:00, it is clear there is a torch-lit, white nationalist march through UVA. The top n-grams for the rest of that day show the escalation to violence and chaos.

Another factor to consider is sustained popularity of particular words or phrases over multiple time intervals. While reappearing words contribute to a story, they also provide a context to recognize when new information is introduced. For example, on August 12, 2017, the day of the planned march, while "trump," "hatred," and "white supremacist" were consistently used, the emergence of "car" between 14:00 and 15:00 clues one into the car ramming attack by Fields around that time. Then, #heatherheyer appears in ~2% of all tweets between 23:00 and midnight before surfacing in the top n-grams over the next hour. From an SA perspective, the first-time appearance of this unfamiliar name in the top n-grams should prompt a closer look into who this person might be, which would reveal that Heather was confirmed as the individual killed in the attack.

Notably, findings also showed a difference in terminology used between UTR sympathizers and counter-protestors, often to describe the same events or people. For example, accounts associating with UTR often referred to Donald Trump as POTUS, whereas counter-protestors used variations of his name (e.g., Trump, Donald, Donald Trump, and D. Trump), perhaps demonstrating a reluctance to accept his position as president. In this manner, decisions to use certain terminology can reveal information about entities in the network regarding their affiliations, perceptions, identity, and positions on various issues.

The final analysis conducted on the Twitter data set concerned the average sentiment of tweets shared by and throughout the network. As seen in Figure 4.4, sentiment within the network vacillates between positive and negative averages until the UVA torch march begins. After that point, sentiment remains consistently negative with additional drops following the commencement of the UTR march on August 12 and the car ramming attack at 13:40. While sentiment scores alone do not provide information about who in the network is unhappy or upset and why, a Twitter network saturated with negatively charged content indicates potential for conflict escalation and confrontation between groups on opposing sides of an issue. In combination with other factors, network sentiment can improve SA and inform decision making for those affected by, or tasked with handling, related events, activities, and communities "on the ground."

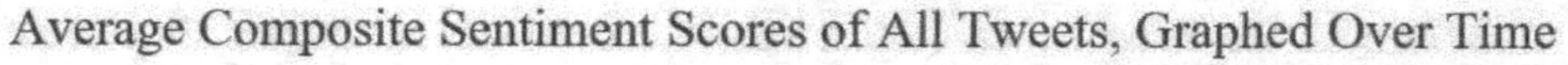

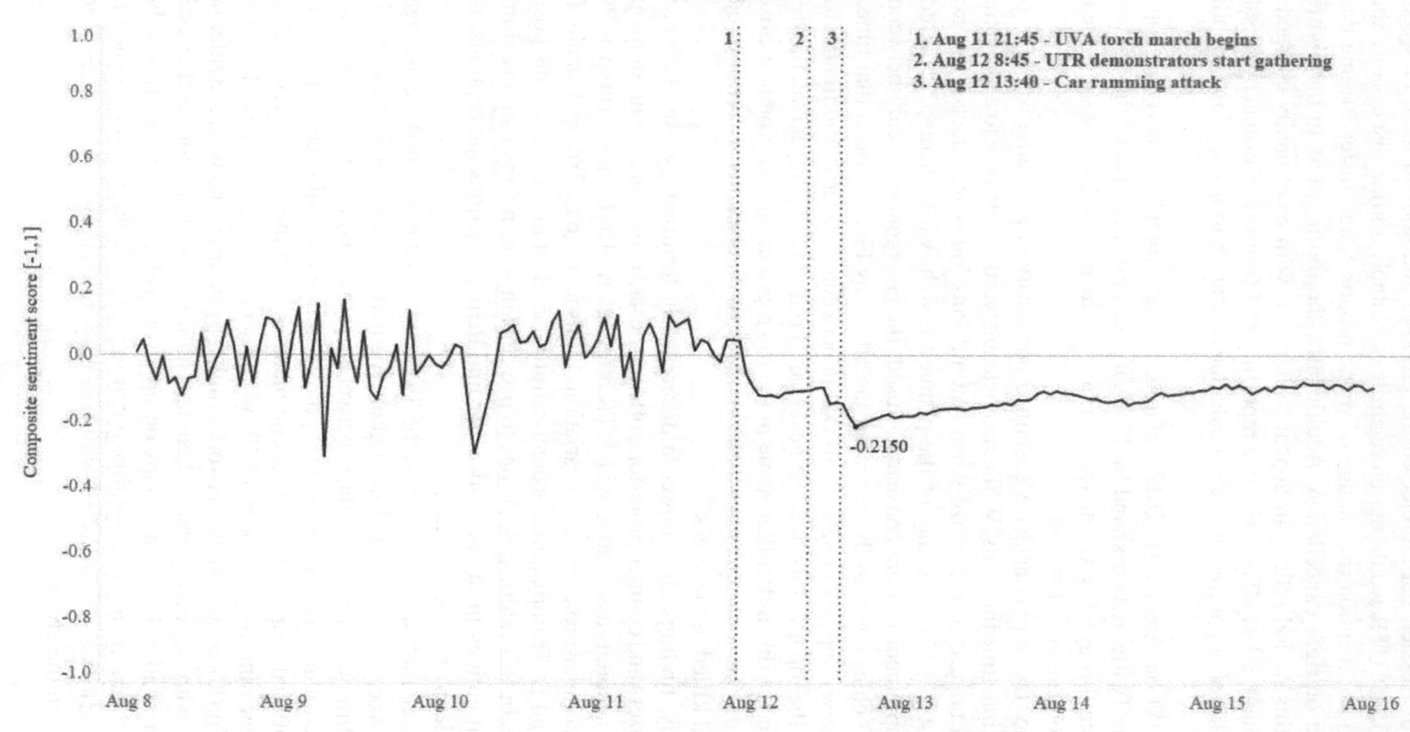

FIGURE 4.4 Average composite sentiment scores of tweets related to the 2017 Unite the Right event in Charlottesville, Virginia, from August 8, 2017 through August 15, 2017 by the hour, overlaid with indicators of notable events.

CONCLUSION

Arno Reuser acknowledged validation of OSINT in today's digital world is essential (Chameleon Associates, 2018). This study set out to provide a framework for such validation by understanding the correlation between what happens in the online and offline spaces. It established a context for fusing digital data from Twitter with information on offline events in Charlottesville in August 2017. It then structured an analytic process that sought to improve situation awareness.

The study supported the theorized calibration of offline situation awareness and action through analysis of online communication within social networks. Patterns in tweet characterization, popular n-grams, and composite sentiment scores offer insight into how messages are created and disseminated, what topics are resonating, and how the network feels about these subjects. In this manner, Twitter can be used to get a sense of a network's position on an issue, establish what is important to the network, and ascertain narratives emerging from the network. Furthermore, it proves that passion is more closely tied to the event and response to what is occurring in real time than pre-event information. Put another way, people are aroused by the emotion of moment and pay little attention prior to an event. All contribute to improving situation awareness and understanding of how Twitter intentionally or unintentionally serves as a mobilizer.

This study also revealed that the operational environment is complex. Those responsible for response and law and order must consider both online and offline data sources to improve situation awareness. Second, while identifying human influencers can help in disrupting events, it is important to recognize the potential of a message to outgrow its creator and to serve as a motivator itself. A strategy of eliminating human influencers has proven to have only marginal success in curtailing messaging. At the operational and tactical levels, disengaging messages' originators or curators is less important than identifying the topics, themes, and banners around which people are coalescing to determine volatility, engagement, and motivation to act. These considerations affect group and crowd behavior.

This study is an initial attempt to provide a framework for understanding the correlation between Twitter activity and real-world events to contribute to situation awareness. It faced some challenges and perhaps created more questions than it answered. Ultimately, this study establishes a foundation for deeper research into the influence of Twitter in facilitating violence and affecting message life cycle, both of which would further improve situation awareness.

BIBLIOGRAPHY

Althoff, T., Jindal, P., & Leskovec, J. (2017). Online actions with offline impact: How online social networks influence online and offline user behavior. In *WSDM 2017 - Proceedings of the 10th ACM international conference on web search and data mining* (pp. 537–546). https://doi.org/10.1145/3018661.3018672

Ashcroft, M., Fisher, A., Kaati, L., Omer, E., & Prucha, N. (2016). Detecting Jihadist messages on Twitter. In *2015 European intelligence and security informatics conference* (pp. 161–164). https://doi.org/10.1109/EISIC.2015.27

Astor, M., Caron, C., & Victor, D. (2017, August 13). A guide to the Charlottesville aftermath. *The New York Times*. Retrieved from https://www.nytimes.com/2017/08/13/us/charlottesville-virginia-overview.html

Berger, J. M., & Morgan, J. (2015). *The ISIS Twitter census: Defining and describing the population of ISIS supporters on Twitter.* Retrieved from http://www.brookings.edu/~/media/research/files/papers/2015/03/isis-Twitter-census-berger-morgan/isis_Twitter_census_berger_morgan.pdf

Berger, J. M., & Strathearn, B. (2013). *Who matters online: Measuring influence, evaluating content and countering violent extremism in online social networks.* London: ICSR.

Bolsover, G., & Howard, P. (2017). Computational propaganda and political big data: Moving toward a more critical research agenda. *Big Data*, *5*(4), 273–276. https://doi.org/10.1089/big.2017.29024.cpr

Bui, T. H. (2016). The influence of social media in Vietnam's elite politics. *Journal of Current Southeast Asian Affairs*, *35*(2), 89–111. https://doi.org/10.1177/186810341603500204

Carter, J. A., Maher, S., & Neumann, P. R. (2014). *#Greenbirds: Measuring importance and influence in Syrian foreign fighter networks.* London: ICSR.

CBSDFW. (2019, June 17). *Accused downtown shooter Brian Clyde: 'A storm is coming'.* Retrieved from https://dfw.cbslocal.com/2019/06/17/active-shooter-shots-fired-federal-building-dallas/

Chameleon Associates. (2018, November 19). *Updating OSINT cycles.* Retrieved from https://chameleonassociates.com/updating-osint-cycles/.

The Economist. (2018, March 28). *How "identitarian" politics is changing Europe.* Retrieved from https://www.economist.com/europe/2018/03/28/how-identitarian-politics-is-changing-europe

Endsley, M. R., & Garland, D. J. (2000). *Situation awareness analysis and measurement* (Vol. 1). Mahwah, NJ: Lawrence Erlbaum Associates Publishers.

Frampton, M., Fisher, A., & Prucha, N. (2017). *The new netwar: Countering extremism online.* London: Policy Exchange. Retrieved from https://policyexchange.org.uk/wp-content/uploads/2017/09/The-New-Netwar-1.pdf

Giesea, J. (2017). *Hacking hearts and minds: How memetic warfare is transforming cyberwar* (Vol. 1). https://doi.org/10.1192/bjp.111.479.1009-a

Hunton & Williams LLP. (2017). Final Report: Independent Review of the 2017 Protest Events in Charlottesville, Virginia.

Hutto, C. J., & Gilbert, E. (2014). VADER: A parsimonious rule-based model for sentiment analysis of social media text. In *Eighth international AAAI conference on weblogs and social media* (pp. 216–225). https://doi.org/10.1210/en.2011-1066

Jason Kessler v. City of Charlottesville and Maurice Jones [Case]: 3:17-cv-00056-GEC.—[s.l.]: US District Court for Western District of Virginia Charlottesville Division, August 11, 2017.

Lind, D. (2014, August 14). Unite the Right, the violent white supremacist rally in Charlottesville, explained. *Vox.* Retrieved from www.vox.com/2017/8/12/16138246/charlottesville-nazi-rally-right-uva.

Littman, J. (2018). *Charlottesville Tweet Ids.* Harvard Dataverse, V1. https://doi.org/10.7910/DVN/DVLJTO

The National Commission on Terrorist Attacks Upon the United States. (2004). *Final report of the National Commission on Terrorist Attacks Upon the United States.* https://doi.org/10.2307/3421510

The Rutherford Institute. (2017, August 10). Rutherford Institute & ACLU sue city for discriminating against controversial "Alt Right" views by blocking protesters' access to park. *Rutherford.org.* Retrieved from www.rutherford.org/publications_resources/on_the_front_lines/rutherford_institute_aclu_sue_city_for_discriminating_against_controversial.

Summers, E. (2017). *The catalog and the hydrator.* Retrieved from Documenting the Now Project. Hydrator Code Repository. https://github.com/DocNow/hydrator.

Summers, E. (2017). *UniteTheRight Tweet IDs.* Retrieved from https://archive.org/details/unitetheright-ids

Tableau Desktop. (2003). Mountain View, CA: Tableau Software.

Taylor, A. (2019) "New Zealand suspect allegedly claimed 'brief contact' with Norwegian mass murderer Anders Breivik". *The Washington Post* March 15. Retrieved from https://www.washingtonpost.com/world/2019/03/15/new-zealand-suspect-allegedly-claimed-brief-contact-with-norwegian-mass-murderer-anders-breivik/

Vidino, L., & Hughes, S. (2015). *ISIS in America: From retweets to Raqqa.* https://doi.org/10.1017/CBO9781107415324.004

Watts, C. (2018). *Messing with the enemy: Surviving in a social media world of hackers, terrorists, russians, and fake news.* New York: HarperCollins Publishers.

Weirman, S., & Alexander, A. (2018). Hyperlinked sympathizers: URLs and the Islamic state. *Studies in Conflict and Terrorism.* https://doi.org/10.1080/1057610X.2018.1457204

Zegart, A., & Morrel, M. (2019). Spies, lies and algorithms: Why U.S. Intelligence agencies must adapt or fail. *Foreign Affairs, 98*(3), 85–96. https://doi.org/10.1002/9781118914472.part1

5 Situation Awareness in Medical Teamwork

Zhan Zhang and Diana Kusunoki

CONTENTS

INTRODUCTION

Research on situation awareness (SA) has been conducted for more than two decades across a variety of domains, including aviation (Gross, 2013), air traffic control (Durso et al., 1998; Endsley, 1999), military operations (Mouloua, Gilson, Kring, & Hancock, 2001), and power plants (Hogg, Folles, Strand-Volden, & Torralba, 1995). Situation awareness has been defined by Endsley (1995) as "the perception of elements in the environment within a volume of time and space, the comprehension of their meaning, and projection of their status in the near future." That is, in each of these dynamic domains, people do more than simply perceive the information and status of various entities within an environment (*perception*); they also integrate and comprehend the information they are perceiving to make sense of what is happening (*comprehension*), and then project the future states of the environment such as what is likely to happen next (*projection*). This observation is well known as the three-level model of SA (Endsley, 1995). The outcomes from this continuous assessment are then used to form the foundation for decision making.

SA has typically been discussed at the level of the individual to denote a person's awareness (Adams, Tenney, & Pew, 1995; Endsley, 1995). However, Endsley (2001) noted that "while SA is essentially a commodity possessed by the individual (as it exists only in the cognition of the human mind), there is nonetheless much to be gained from examining SA as it exists within teams and between teams that are involved in achieving a common goal." In collaborative work, individuals need to understand the status of the system with which they are working, while being aware of what other individuals or teams are doing.

Along the same lines, several workplace studies have examined the key roles of situation awareness in teamwork, showing how collaborators present, align, and integrate their communication and activities by maintaining an awareness of each other's work and intentions (e.g. Berndtsson & Normark, 1999; Heath & Luff, 1992; Hutchins, 1995). Situation awareness is especially critical in medical teamwork. The increasing specialization of medical knowledge and services, as well as the distributed nature of hospital work, have led to a large number of studies examining the ways in which situation awareness is maintained in medical contexts and how to support awareness in collaborative medical work (e.g. Bardram, Hansen, & Soegaard, 2006; Bossen, 2002; Reddy et al., 2009; Svensson, Heath, & Luff, 2007; Zhang & Sarcevic, 2015).

In this chapter, we will first review the research of situation awareness in medical settings that feature collocated, distributed, synchronous, and asynchronous collaborative work. This review highlights the essential facets of situation awareness and the supporting technologies that have been proposed and discussed in prior work. Then we will use trauma care as an example domain to provide a micro-level, domain-specific perspective on what these facets of situation awareness mean in the context of ad hoc, interdisciplinary, and time-critical medical teamwork. Lastly, we draw upon activity theory (AT) to present a conceptual model of situation awareness that can be used to understand time-critical and interdisciplinary teamwork, and to identify the technological needs for supporting different kinds of awareness.

FACETS OF SITUATION AWARENESS IN MEDICAL TEAMWORK

Medical professionals need to maintain awareness of different aspects of teamwork—some are related to the team and others to tasks or the care process. This section reviews the healthcare studies that describe the multiple facets of situation awareness in medical teamwork, including social, temporal, spatial, activity, and process awareness. Differentiating between these facets is important because it can provide researchers with a framework with which to examine situation awareness in medical teamwork and develop guidelines for providing tailored awareness support.

Social Awareness

Social awareness has been described as knowing who is around, where and how far, and what is their current status and availability (Bardram et al., 2006; Carroll, Neale, Isenhour, Rosson, & McCrickard, 2003; Kusunoki, Sarcevic, Zhang, & Yala, 2015; Prinz, 1999). Being aware of each other's work context is a core mechanism for initiating proper conversation and engaging in cooperative efforts between co-workers. Maintaining social awareness within a team of clinicians has proven to be central to the coordination of work in hospitals (Bardram et al., 2006). For example, nurses spend time on maintaining awareness of where relevant clinicians are, their current work activities, and their schedules. When a nurse needs to contact a clinician, this awareness helps him or her decide who to contact, when, where, and how. Furthermore, clinicians have different skillsets and levels of experience. Nurses who have worked for over 20 years at a hospital ward will know how to approach specific diseases or work protocols faster than clinicians with less experience. In this sense, clinicians need to be aware of not only the whereabouts and statuses of co-workers, but they also need to learn who has the competency and experience to perform certain tasks (Bossen, 2002). For example, chemotherapy delivery requires at least two nurses specifically trained in this process who need to collaborate on a series of tasks, such as confirming the patient's identity, calculating medication doses, and administering the medications. To accomplish this task, a nurse therefore needs to know who else is qualified to do this and then locate that nurse in the hospital ward (Bossen, 2002).

Research on social awareness in medical settings has mostly focused on collocated clinicians or clinicians working in close proximity to receive visual cues about events taking place (e.g. Heath, Svensson, Hindmarsh, Luff, & Vom Lehn, 2002; Zhang & Sarcevic, 2015). As clinicians are collocated and share the same work context, they can take advantage of working side-by-side to maintain a mental picture of each other's status and availability. Collocated clinicians can achieve such awareness through two essential mechanisms: *displaying* and *monitoring* (Bardram & Hansen,

2004; Heath et al., 2002). Bardram and Hansen (2004, p. 193) describes the term *displaying* as "implicit or explicit signals a given actor uses to show specific aspects of his or her current situation, which could be useful or relevant for the other actors in the context." In the study of a collocated time-critical medical setting, Zhang and Sarcevic (2015) noted that trauma resuscitation members leveraged different mechanisms and embodied actions (e.g., body movement, gesture, eye gaze, etc.) to implicitly display their current status (i.e., readiness to perform more tasks). On the other hand, the notion of *monitoring* requires team members to continuously observe the work status of others to tacitly and unobtrusively align their own actions (Bardram & Hansen, 2004; Zhang & Sarcevic, 2015). *Monitoring* relates to *reading a scene*, both of which involve assembling knowledge "through juxtaposition and interpretation of verbal reports, visual images, and various forms of text in real time" (Suchman, 1997, p. 49).

In contrast, it is very challenging for clinicians to maintain the same level of social awareness when they are distributed across greater distances, such as a large hospital, and therefore need to rely on various means and mechanisms to construct awareness of each other's work context. In a study of hospital patient wards, Bossen (2002) illustrated the use of symbols and signs to reveal the activities, the status of the work, and the location of clinicians. For instance, during the daily round when physicians visit patients' rooms, a colored sheet of paper is put up with a peg outside the room. By doing so, other clinicians would know where the physician may be located.

Clinicians also used telephones and pagers to look for other hospital staff. However, the extensive use of communication means between distributed co-workers could lead to unintended interruptions and disturbances (Bardram & Hansen, 2004; Bardram et al., 2006). For example, a nurse wants to contact the physician to inquire about a patient while the physician is discussing radiology images with an expert at the radiology department. In this case, the physician might be disturbed by the nurse calling. Constructing social awareness is thus also important for minimizing interruptions when people engage in cooperative work.

Temporal Awareness

Temporal awareness is crucial to collaboration in medical work as it provides clinicians with an awareness about past, present, and future activities (Bardram, 2000; Reddy, Dourish, & Pratt, 2006). A surgeon scheduled for an operation would plan their other activities around it accordingly, such as when to perform regular ward rounds, prepare the patient for operation, and conduct brief meetings with the clinicians attending the operation (Bardram & Hansen, 2010a).

Seminal studies of coordination in medical work have looked closely at temporal awareness, including temporal coordination and temporal rhythms. Temporal coordination is defined as "an activity with the objective to ensure that the distributed actions realizing a collaborative activity takes place at an appropriate time, both in relation to the activity's other actions and in relation to other relevant sets of neighbor activities" (Bardram, 2000, p. 163). Based on in-depth studies of the socio-temporal aspects of coordinative work at a surgical department, Bardram and colleagues

(2006, 2010a) discussed the importance of temporal coordination for activities such as scheduling patient care, synchronizing actions, and allocating time for working together, and how temporal coordination is mediated by temporal coordination artifacts (e.g., work schedule) and is shaped according to the temporal conditions of the collaborative activity (Bardram, 2000; Bardram et al., 2006).

Temporal rhythms were first discussed in the classic study by Zerubavel, where the rhythmic structures of social organization in hospital life were characterized by five major social cycles: the year, the rotation, the week, the day, and the "duty period" (Zerubavel, 1979). Reddy and colleagues (Reddy and Dourish, 2002; Reddy et al., 2006) then noted the importance of temporal rhythms in providing clinicians in a surgical intensive care unit (ICU) with a resource for seeking, providing, and managing information in the course of their work. They found that temporal rhythms ensure orderliness of the ICU work by orienting ICU clinicians towards likely future activities and information needs. Such rhythms include large-scale rhythms (e.g., nursing shifts, rounds) and fine-grained rhythms (e.g., lab results, medication administration). Medications are usually given on a known schedule, therefore nurses can arrange their activities around this schedule.

Spatial Awareness

Spatial awareness has been defined as knowing the activities taking place in a specific space and how people are interacting with the space itself (Bardram et al., 2006; Gutwin & Greenberg, 2002). Spatial awareness is typically described in contexts where actors or teams are distributed to varying degrees (Kusunoki et al., 2015). It is therefore vitally important to enable those who are not physically present at a given location to maintain awareness of what is happening there. The hospital is a setting where coordination takes place in distributed locations, with clinicians, patients, and equipment moving throughout different departments that are usually highly specialized (Bossen, 2002; Scupelli, Xiao, Fussell, Kiesler, & Gross, 2010; Tellioğlu & Wagner, 2001). Clinicians thus spend a lot of effort on maintaining awareness of what is happening in a particular location (e.g., patient room, operating room).

In studies of operating rooms (OR), Bardram and his colleagues (Bardram et al., 2006) described various means and mechanisms used in maintaining spatial awareness of the OR suite, such as the OR whiteboard showing the status of an operation. They highlighted the features of spatial awareness that are of the utmost importance in the coordination of OR work and need to be made available all the time, including the type and status of operation taking place in a given operating room, the people in the room, and any contingencies (e.g., delays). In a similar context, Scupelli et al. (2010) examined another aspect of spatial awareness—how the physical environment of surgical suites support or fail to support collaboration between clinicians. They argue that these architectural dimensions of hospitals can significantly impact information access and interpersonal interactions. That is, barriers such as corridors, stations, walls, and stairways reduce clinicians' visual and auditory access between workspaces, ultimately reducing opportunities for collaboration and maintaining spatial awareness. They suggest that artifacts (e.g., schedule whiteboard) should be placed in an appropriate location so that clinicians can easily encounter these artifacts to find information during the course of their work activities and be aware of the activities taking place in the surgical suites. Furthermore, placing such collaborative artifacts where surgeons, nurses, anesthesiologists, and other clinicians are likely to pass by at the same time can better facilitate conversation initiation and teamwork.

Activity Awareness

Activity awareness is an awareness of particular activities and the context in which they take place (Bardram & Hansen, 2010b; Convertino et al., 2008). As Carroll, Rossen, Farrooq, and Xiao (2009) noted: "collaborators need be aware of a whole, shared activity as complex, socially and culturally embedded endeavour, organized in dynamic hierarchies, and not merely aware of the synchronous and easily noticeable aspects of the activities" (pp. 166).

Studies of activity awareness in medical environments describe this concept as knowing what other clinicians did, what they are doing, or what needs to be done (Cabitza, Simone, & Zorzato, 2007; Dourish & Bellotti, 1992; Prinz, 1999). As medical teams are typically comprised of multiple professions (e.g., nurses and physicians) with different job titles and responsibilities, clinicians need to keep track of what others are doing so that they can determine what they need to do to achieve seamless coordination. For example, in the study of coordination in trauma resuscitation, Zhang and Sarcevic (2015) illustrated how trauma team members rely on their ability to monitor each other's activities, assess the relevance of those activities to their own work, and anticipate their next move. Through examining the use of a particular artifact (clinical pathway) at a neonatal intensive care unit (NICU), Cabitza et al. (2007) identified four facets of activity awareness that need technology support: accounting awareness, reminding awareness, coordinative awareness, and enabling and inhibiting awareness. In the operation ward setting, Bardram et al. (2006, Bardram & Hansen, 2010a, 2010b) showed that surgeons have to manage and keep track of concurrent activities in their daily workflow. They used different methods to maintain such awareness and adjust their work accordingly. Surgeons may phone other departments regarding the status of certain requests (e.g., if and when the lab test results will be ready), or read the work plans to get a sense for other clinicians' activities at the moment. Bardram et al. (2006, 2010a, 2010b) suggested that there are two general approaches in supporting the construction of activity awareness in distributed hospital settings: (1) presenting general information about an activity to the entire team and (2) notifying specific inquirers about the progress of that activity.

Process Awareness

Process awareness has been defined as knowing the general sequence of tasks and current status of the process (Cabitza, Simone, & Zorzato, 2009). Few studies have looked into the importance of process awareness in collaborative work. Cabitza et al. (2007) highlighted the importance of developing a computational model to integrate clinical and organizational processes. One of the practical applications of this work is the concept of a computational system that correlates clinical processes with the data structures of medical records (clinical pathway) in order to achieve better coordination in daily clinical practices. In a later study, Cabitza et al. (2009) emphasized the key role of a "process-aware" documentation system in supporting "articulation work" (Schmidt & Bannon, 1992; Strauss, 1988). Clinicians can use this system to assess and set the current state of the clinical process, consult the process history, and align their practices to a specific protocol, procedure, or process.

Summary

The majority of situation awareness research in medical settings have identified important facets of awareness that are essential in collaborative work. These facets of awareness have been described in detail with regard to collocated, distributed, synchronous, and asynchronous contexts. It is worth noting that these facets of awareness are correlated and interdependent to some extent. While spatial awareness differs from social awareness, in that spatial awareness focuses on how people interact with each other inside a physical space, they are both concerned with people, as in who is around and available for collaboration. Similarly, there is a link between the procedural or process aspects of work practice and the activities of which the process usually consists. That is, both process awareness and activity awareness require the knowledge of contextual information about what past, ongoing, and future activities are included in a process. The correlation between different facets of awareness suggests that awareness technologies should be designed to support a set of interdependent facets of situation awareness rather than only one. In the following section, we will address the awareness research about designing and developing technical systems to support different facets of situation awareness in dynamic medical settings.

SITUATION AWARENESS SUPPORTING TECHNOLOGY IN MEDICAL SETTINGS

The technology for situation awareness support needs to capture, manage, and distribute awareness information. This section discusses examples of technologies that have been designed and developed in prior work to support different aspects of situation awareness in medical settings. Here we focus on two types of technology support for awareness: (1) awareness information infrastructure and (2) awareness information presentation.

Awareness Information Infrastructure

Awareness information infrastructures have been built to provide clinicians with awareness support through different systems, platforms, sensors, and applications. They can automatically capture, process, and store information to support awareness from various data sources, then distribute this information to interested users in diverse formats such as digital displays, mobile applications, and computer desktops.

We review the QoS DREAM framework and the AWARE architecture to exemplify this concept.

Mitchell, Spiteri, Bates, and Coulouris (2000) built a "follow-me" multimedia application for hospitals on top of the QoS DREAM architecture. This architecture integrates a multimedia framework with an event-based notification system, which is aimed at providing context-sensitive communication channels and awareness information to distributed clinicians. An important feature of this system is that it can automatically detect where a clinician is located and send the requested information to that specific location. A number of capacitive touchscreen terminals have been deployed throughout the hospital. Clinicians can walk up to a terminal and request a video or audio call with other staff in the hospital. It is not necessary to know where the other person is currently located because the system is aware of the location of all staff members. The system also allows clinicians to keep track of equipment and patient names, locations, statuses, and data. Authorized clinicians could use event-based notifications to monitor patient data (e.g., test results) of interest that can then be distributed to a nearby terminal for them to retrieve.

In a more recent example, Bardram and colleagues (Bardram & Hansen, 2004; Bardram et al., 2006) worked closely with clinicians to develop the AWARE architecture, which focuses on providing a variety of awareness information in a surgical department. This architecture addresses four dimensions of context-based awareness: social, spatial, temporal, and activity awareness. The entire system consists of different components, including the integrated AwarePhone and AwareMedia applications, and the underlying architecture. The AwarePhone application is a smartphone application with a contact list that supports basic text messaging. This application was designed to reduce interruptions by displaying the work context of each user (e.g., personal status, activity, and location) directly on the contact list for clinicians to keep in mind before initiating a call to their colleague. Given the great mobility of clinicians, they may not have easy and timely access to office equipment such as desktop or laptop computers. The AwarePhone system supports clinicians who are always moving to obtain awareness of work progress. The AwareMedia system consists of public, interactive displays installed in various places of the hospital, providing an overview of the coordination center and operating rooms, clinicians' status and location, and surgery schedules. These applications combined with mobile phone capabilities helped improve coordination and awareness among surgical team members.

Overall, awareness information infrastructures were deployed in various physical locations, providing awareness support beyond individual applications. The infrastructures were based on data captured by sensors about people, locations, and activities, and notifications to distribute awareness information to interested and authorized clinicians. The QoS DREAM system was able to locate clinicians and enabled conference calls and other data to follow the user while moving. This system demonstrated the feasibility of collecting contextual information about other actors in a system, that can be presented to the user to support social awareness. The AWARE architecture extended this concept by supporting more than one type of awareness. In addition, it highlighted the importance of engaging end users (i.e., clinicians) in the design and evaluation process.

Awareness Information Presentation

Presentation of awareness information is an integral part of supporting awareness and can be presented in different formats on various devices. Here we focus on two commonly used technology solutions in medical settings for presenting awareness information: information displays and mobile applications.

One of the most effective ways to present awareness information is through shared information displays (Wallace, Scott, Lai, & Jajalla, 2011). Integration and display of large amounts of contextual information about patient status and team tasks have been used for augmenting communication, work coordination, and awareness in a variety of medical settings, including critical care units (Wilson, Galliers, & Fone, 2006), operating rooms (Drews & Westenskow, 2006; Lai, Spitz, & Brzezinski, 2006; Levine et al., 2005; Parush et al., 2011), and emergency departments (France et al., 2005; Wears, Perry, Wilson, Galliers, & Fone, 2007). These shared displays and status boards have been shown to support both collocated and distributed work by facilitating task coordination, resource planning, communication, and problem solving. For instance, Parush et al. (2011) examined the situation-related communications in the cardiac OR and proposed a design concept of a digital display to augment the OR team's situation awareness. The display includes several visual components related to patients, procedures, team information, and vital signs. An overall timeline visualizes the phases of surgery, highlighting key events and actions. They argued that "successful design of information sharing applications should be based on an understanding of communication patterns among healthcare workers and the nature and timing of sharing situation-related information" (p. 483). Similarly, Lai et al. (2006) and Levine et al. (2005) emphasized the integration and presentation of both persistent and evolving information from disparate sources in the OR to enhance situation awareness. Persistent information includes patient demographics, diagnoses, procedures, and staffing information, while dynamic information includes the past, present, and future stages of an operation. They suggested that shared displays can facilitate team orientation throughout an operation, especially during staff changes.

Pervasive devices and mobile applications have become essential for collaborating clinicians that are both "mobile" and distributed. Favela and Martinez-Garcia (2003) highlighted the need for context-aware communication in hospitals by providing information about team members, such as location, role, and status, to other hospital staff. They proposed a context-aware instant messaging (IM) application running on portable digital assistant (PDA) devices for clinicians to access and distribute awareness information as they moving around the hospital. This application allows clinicians to distribute a message by indicating the location or work context instead of sending it to specific recipients. In a recent research effort, Lane, Sandberg, and Rothman (2012) developed and implemented a mobile application to support situation awareness in an operative environment. This mobile application allows anesthesiologists to know the status of their patient and which patient needs immediate attention by providing access to a visual map of the flow of patients throughout the operative suite. In particular, the application enables anesthesiologists to maintain an awareness of the status of their patients in up to four locations simultaneously.

Overall, information displays and mobile applications are common solutions used to present awareness information. The solutions described in this section are just a few of many examples. Mobile applications are particularly suited to situations where clinicians are constantly changing location and collaborating over distance. Information displays, on the other hand, are typically developed for collocated teams, such as an OR team, but can also be ubiquitously deployed in a hospital, allowing clinicians to access awareness information while they are on the move.

SITUATION AWARENESS IN AD HOC, INTERDISCIPLINARY, TIME-CRITICAL MEDICAL TEAMWORK

Prior research on situation awareness in clinical settings has mostly focused on collaborative work spanning across long-term trajectories that range from hours and weeks to even months. Few studies directly examine the details of awareness in ad hoc, fast-paced teamwork, with only a few exceptions (Strater, Cuevas, Connors, Ungvarsky, & Endsley, 2008; Strater et al., 2009). In addition, previous studies focused on either a collocated medical setting or a distributed environment, with limited attention paid to a medical context with both intra- and inter-team collaborative work (Endsley & Jones, 2001). In this section, we illustrate what situation awareness means within the context of an ad hoc, time- and safety-critical medical setting that features both collocated and distributed teamwork. The purpose of our research is not to define awareness or identify new facets of situation awareness. Instead, we build on the existing understanding of situation awareness in other medical contexts and offer a micro-level, domain-specific perspective on these facets in the context of ad hoc, multidisciplinary medical teamwork.

Problem Domain: Trauma Care

Trauma care is a fast-paced and dynamic process that addresses life-threatening injuries. The first hour after injury—the "Golden Hour"—is a critical period that greatly determines a patient's chances of survival (Spanjersberg, Bergs, Mushkudiani, Klimek, & Schipper, 2009). It is therefore essential that emergency medical teams efficiently and seamlessly coordinate to treat patients. The trauma care process typically involves several teams from different disciplines (Figure 5.1). The Emergency Medical Services (EMS) team provides preliminary care in the field, then transports patients to the nearest hospital. The Emergency Communication and Information Center (ECIC) is a hospital facility where a team of communication specialists coordinate patient care and transport. The ECIC is the first point of contact for anyone transporting or sending a patient to the hospital. The Emergency Department (ED) team provides emergency diagnoses and treatments for a wide range of illnesses and injuries. Critically injured patients are treated by the trauma team at a dedicated resuscitation room (trauma bay) in the ED (Burd & Elliot, 2011; Ludwig et al., 2010). Trauma teams typically consist of surgical attending physicians, emergency medical physicians, physician surveyors, respiratory therapists, anesthesiologists, and nurses (Figure 5.2). Each team member has a specific role with a set of well-defined responsibilities (Burd & Elliot, 2011; Ludwig et al., 2010). Surgical attending physicians

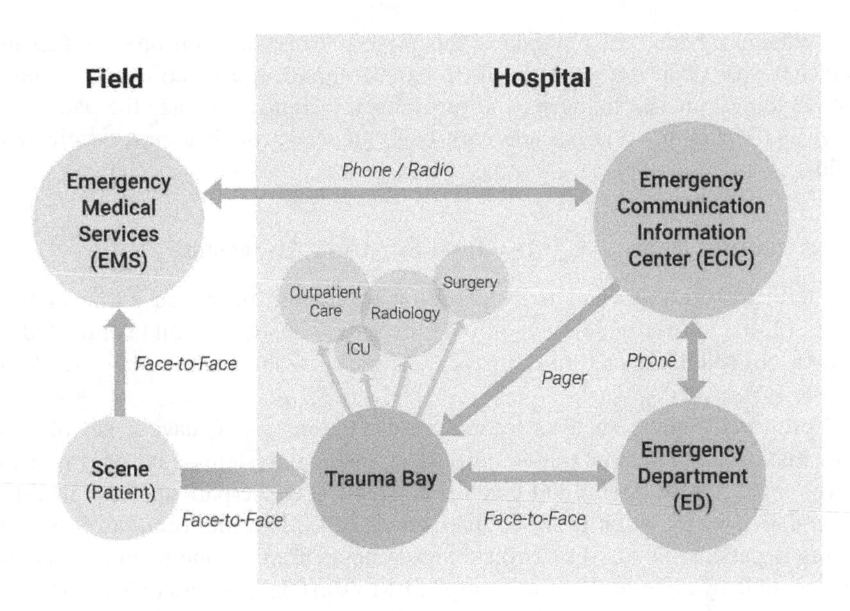

FIGURE 5.1 Trauma care collaboration and communication process.

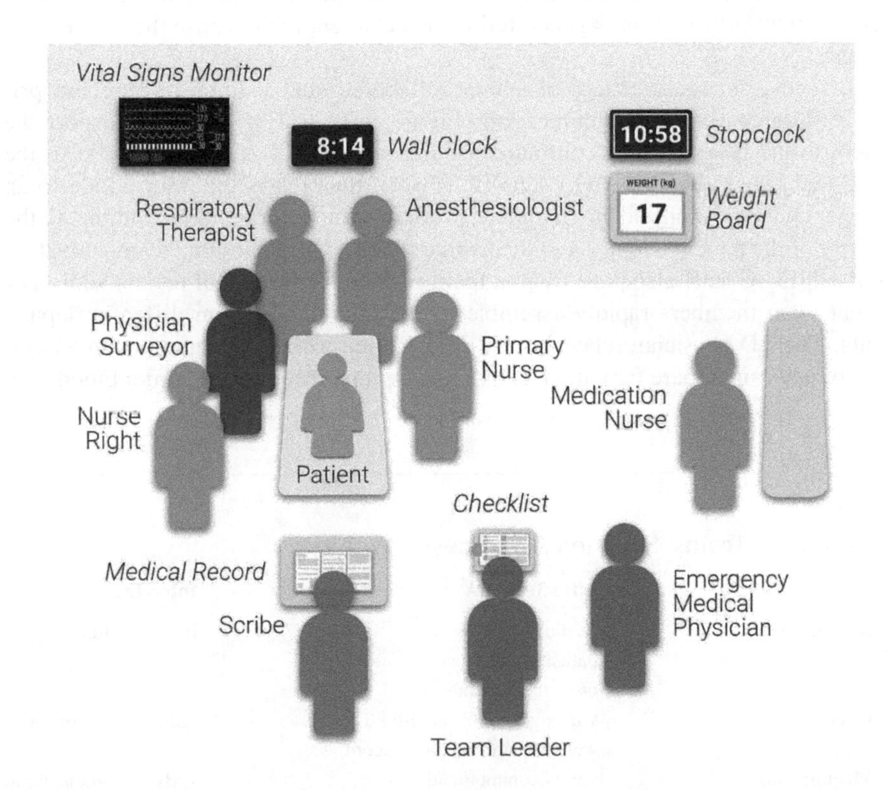

FIGURE 5.2 Collocated trauma team roles, information tools, and information technology positions relative to the patient.

and emergency medical physicians supervise the resuscitation process and make decisions; physician surveyors perform hands-on patient evaluations and announce their findings; anesthesiologists and respiratory therapists manage the patient's airway; and nurses provide bedside care (e.g., preparing medications, administering fluids).

Constructing Inter- and Intra-Team Situation Awareness

We will use the model of team situation awareness proposed by Endsley and Jones (2001) to briefly describe how team SA is constructed and supported both within collocated teams and across distributed teams in trauma care domain (Table 5.1).

Collocated teams, such as trauma resuscitation teams, have a set of shared information (e.g., patient status, outputs of medical systems, and who is present in the resuscitation room) and goals (i.e., stabilizing the patient). In contrast, the only shared information between distributed EMS, ED, and trauma resuscitation teams is patient status. The artifacts and systems used by one team are generally inaccessible by other teams; for example, EMS staff have access to documents they used to record patient information in the field while the receiving hospital teams (e.g., ED physicians, trauma resuscitation members) do not. Sharing the context in which patient information is generated is also challenging because the visual cues are lost.

Currently, emergency medical teams collaborate and maintain awareness primarily through verbal communication (Figure 5.1). As EMS teams transport the patient to the hospital, they communicate information collected in the field to the ECIC team via radio or cellular. The ECIC staff then relays the information to an emergency medical physician via phone, who uses this information to anticipate the severity of the patient's injury and determine the level of the trauma team activation required. ECIC staff also page trauma team members with a brief patient summary so that team members rapidly assemble in the trauma bay from different departments. The ED physician relays known information while the trauma team assembles so they can prepare for patient arrival (e.g., prepare equipment, order blood, and

TABLE 5.1
Trauma Care Teams' Situation Awareness

	Intra-Team SA	Inter-Team SA
SA Requirements	medical systems patient status other team members	patient status
SA Devices	vital signs monitor, checklist, weight board, medical record	radio/cellular phone
SA Mechanisms	verbal communication	verbal communication
SA Processes	stabilization protocol	stabilization protocol

summon specialists). When the patient arrives, an EMS team member announces a summary of patient injuries and treatments administered en route. Once patient hand-off is completed, the trauma team begins the resuscitation process of assessing and treating the patient, typically lasting between 10 to 20 minutes. Once the patient is stabilized, they are transferred to another unit for further treatment (e.g., surgery, radiology, ICU, outpatient care).

During the resuscitation process, trauma team members have the advantage of being collocated to not only share information directly face-to-face but to also overhear conversations and tacitly monitor each other's work—a widely used mechanism for maintaining awareness in dynamic teamwork settings (Heath & Luff, 1992). Each role produces different types of information that can be announced to the team to be consumed. Even when tasks are discussed among just a few people, team members can benefit from overhearing information about the work status of others to plan and align their own actions (Zhang & Sarcevic, 2015).

Trauma team members also use information tools and technologies to maintain awareness. The most common technologies include the vital signs monitor, temporal artifacts, paper-based medical records, and whiteboards (Figure 5.2). The vital signs monitor is an essential information technology that provides feedback to clinicians about the patient's status. Temporal artifacts typically include a wall clock, stop clock, and timers to help trauma team members manage their time (Kusunoki & Sarcevic, 2015). There are also paper-based trauma medical records used to document the process manually and checklists to ensure that all process steps are completed. Whiteboards are often used to display critical patient information such as age, weight, and mechanism of injury. These basic pieces of information can support shared situation awareness among collocated trauma team members.

Trauma care teams must explicitly share and request information in order to construct and sustain awareness in this time-critical environment, but many factors can complicate their ability to do so. Trauma team members come from different departments and may not necessarily know each other, making the team less efficient in establishing common ground (Lee, Tang, Park, & Chen, 2012; Majchrzak, More, & Faraj, 2012; Sarcevic, Marsic, Waterhouse, Stockwell, & Burd, 2011). Furthermore, the distributed nature of the trauma care process and the use of traditional telecommunication channels (e.g., radio, telephone) pose significant challenges in constructing precise awareness of the patient's status among the different emergency medical teams (Zhang, Sarcevic, & Bossen, 2017).

To address these challenges, we conducted multiple field studies in an urban teaching hospital and trauma center over the course of six years to understand how these interdisciplinary teams coordinate work both within and across teams and construct situation awareness in fast-paced and dynamic medical contexts in order to design meaningful awareness support. We conducted interviews, observations, video reviews, artifact analysis, and participatory design workshops to gain clinicians perspectives on the challenges they face in maintaining situation awareness during trauma care. More details, including research methodologies and findings, have been reported in our previous work (Kusunoki & Sarcevic, 2015; Kusunoki et al., 2014; Kusunoki et al., 2015; Zhang & Sarcevic, 2015; Zhang et al., 2017).

In the next section, we highlight the main aspects of situation awareness that emergency medical teams need to manage throughout the trauma care process. The intent is to use the context of the trauma care domain to anchor the discussion of our theoretical framework for understanding situation awareness that we present later.

Insights into Situation Awareness from the Perspective of Interdisciplinary Teams in Emergency Medical Settings

We adapted the facets of situation awareness characterized in the literature to understand the micro-level awareness needs of clinicians working in a time-critical environment. The four most prominent aspects of situation awareness that emergency care teams manage during their collaborative work when treating severely ill patients are: (1) team member awareness (i.e., social and spatial awareness), (2) elapsed time awareness (i.e., temporal awareness), (3) teamwork-oriented and object-driven task awareness (i.e., activity awareness), and (4) overall progress awareness (i.e., process awareness). Below we describe these four aspects of situation awareness in further detail.

Social and Spatial Awareness—Team Member Awareness

Our work revealed a clear distinction between constructing social and spatial awareness both *inside* and *outside* the trauma bay. Resuscitation team members are colocated and coordinate their work synchronously so it is easier to visually monitor the activities in which each person is engaged. They are mainly concerned with who is in the room at the moment and which roles are missing to determine how to compensate. However, unlike surgical (Bardram et al., 2006) or ICU teams (Cabitza & Simone, 2007) that have the opportunity to develop intimate knowledge of each other's work after working together on a regular basis, resuscitation teams form ad hoc—with clinicians coming from different disciplines and not necessarily knowing each other—requiring intense articulation work to construct awareness of who is responsible for what and which roles are present or missing (Bossen, 2002). For example, clinicians introduce themselves before starting resuscitations. They also verbally announce what they are working on, what they completed, and what they plan to do next. To some extent, these mechanisms alleviate the potential risk of establishing awareness caused by the ad hoc nature of resuscitation teams, i.e., clinicians have no existing connections and relationships on which to build common ground.

The information needed for achieving social and spatial awareness outside the trauma bay is mainly about the location and status of the EMS team transporting the patient and consulting specialists. As emergency medical teams involved in trauma care are distributed, locating and assembling clinicians is challenging. It is especially challenging when trauma team leaders decide to call in a specialist to consult based on the patient's injury. Estimating how long it will take for the specialist to arrive is difficult, often requiring extra effort to make follow up calls to determine their status. The ECIC and trauma teams at the receiving hospital also need to know the location of the EMS team to estimate patient arrival time so that trauma team members can assemble and prepare for the patient in a timely manner.

Social and spatial awareness in the context of ad hoc, inter-, and intra-emergency medical teamwork can be conceptualized as *team member awareness*—that is, *knowing who is present, absent, or en route, who is responsible for certain tasks, who is bringing in the patient, and where a particular clinician is located.*

Temporal Awareness—Elapsed and Estimated Time Awareness

Temporal awareness in prior work is mainly discussed in longer-term medical contexts that span hours, days, or months (Bardram, 2000; Bardram et al., 2006; Reddy & Dourish, 2002; Reddy et al., 2006). In contrast, emergency medical teams coordinate work under greater time pressure and the awareness of past, present, and future activities is situated within a significantly condensed time frame (i.e., minutes and even seconds), where elapsed time is essential for coordinating sequentially dependent tasks. As a key example, before intubating the patient, anesthesiologists and nurses must coordinate closely to prepare and administer medications for intubation. Usually the nurse is responsible for preparing and administering the medication, but the anesthesiologist must know the exact moment that specific medications are administered so they can intubate the patient within the very limited time frame (i.e., three minutes) before the medication loses its efficacy. This task interdependency requires that not only the tasks be completed in the correct order, but also in a timely manner. Elapsed time, combined with changes in patient status after performing certain interventions (e.g., intubation), can help clinicians determine if an intervention is effective. Finally, elapsed time awareness is also important because it allows clinicians to estimate how much time they have left to prepare for and perform certain procedures.

Therefore, temporal awareness in emergency medical settings can be considered as *knowing the estimated time of the patient's arrival, estimated time to complete tasks, time elapsed since interventions or certain tasks, and time elapsed since changes in patient status.*

Activity Awareness—Teamwork-Oriented and Object-Driven Task Awareness

Similar to team member awareness, there is also a clear distinction between maintaining awareness of activities taking place both inside and outside the trauma bay. Clinicians in the trauma bay can visually monitor the status of ongoing tasks and overhear conversations to obtain information. In contrast, communicating the context in which patient information is produced across distributed emergency medical teams is challenging because the tacit, visual aspects of communication are missing. Teams still use analog channels such as radio and telephone to send patient information, limiting the amount and types of awareness information that can be conveyed across teams.

In the context of trauma care, emergency medical teams must construct and sustain *teamwork-oriented* awareness as these teams perform interdependent medical activities. Critically-injured patients are evaluated and treated at different times and locations by different teams. After patient hand-off, trauma teams need to know what treatments and physical examinations have been performed by EMS teams in the field and en route to the hospital, so they can determine the necessity and priority of further actions.

Emergency medical teams also need to maintain *object-driven task awareness.* The patient, which can be conceptualized as an object, is a critical source of information in medical work. Clinicians must maintain awareness of the patient's status to see how the patient is responding to treatments and interventions. They must also remember important patient characteristics including sex, age, and weight (often written on a whiteboard) to determine the proper medications and dosages. Contextual patient information such as the mechanism of injury, pre-hospital interventions, and current patient status allows trauma teams to better anticipate the severity of the patient's injury, make informed decisions, and evaluate the effectiveness of each treatment. Furthermore, the patient's status and vital signs can change dramatically throughout the course of trauma care, requiring emergency medical teams to frequently check the vital signs monitor and dynamically adapt their care. Patient status en route to the hospital can be drastically different from when the patient arrives at the hospital. Such patient status changes highlight the fact that awareness in emergency medical settings is dynamic and ongoing.

Activity awareness in this context can therefore be defined as *teamwork-oriented and object-driven task awareness*—that is, *knowing the status and progress of individual tasks, interdependency among tasks, and contextual information about the patient.*

Process Awareness—Overall Progress Awareness

In emergency medical settings, clinicians must aggregate their awareness of activities, team members, and elapsed time to gain an overview of the progress made in the trauma care process. All emergency care teams follow established protocols such as the advanced trauma life support (ATLS) protocol to plan and dynamically manage their tasks, which in turn, helps construct and maintain their overall awareness of team's progress. Patient status can change at any moment so clinicians need to frequently check to see if they need to return to a step. For example, the trauma resuscitation team may be working on the patient's circulation (third step of ATLS protocol), but the patient's airway suddenly deteriorates, and they must collectively shift the focus of their work to re-address the airway (first step). In addition, trauma team members may arrive late to the trauma bay and need to synchronize their awareness of the tasks and overall progress of the trauma care process with the rest of the team. Trauma care teams, especially clinicians in leadership roles, need to maintain awareness of *what procedures have been performed, what protocol step the team is currently working on, and what still needs to completed.*

Summary

We adapted the facets of awareness characterized in the literature to understand the micro-level awareness needs of clinicians working in a time-critical environment featuring both collocated and distributed teamwork. These facets of awareness are essential for collaborative work across the trauma care process. *Team member awareness* emphasizes how different roles and occupations of fast-response teams coordinate tasks over a short time span. *Teamwork-oriented and object-driven task awareness* centers on the feedback, interdependencies, and individual progress of tasks of the team—all driven by the patient's status. *Elapsed and estimated time awareness* focuses on estimating the amount of time that has passed since or until major events, including the patient's arrival, interventions, critical tasks, or changes in patient status. Finally, *overall progress awareness* centers on the general status of the patient and teamwork taking place across the whole trauma care process. Overall progress awareness is comprised of the other three overlapping, interdependent facets of awareness (Figure 5.3).

Implications for Supporting Awareness and Teamwork in Critical Medical Contexts

This trauma care teamwork case shows that medical professionals have high cognitive load as they need to maintain high-level awareness of all the tasks completed, in progress, and pending while treating and stabilizing critically injured patients under time pressure. In such fast-paced, ad hoc contexts, the team

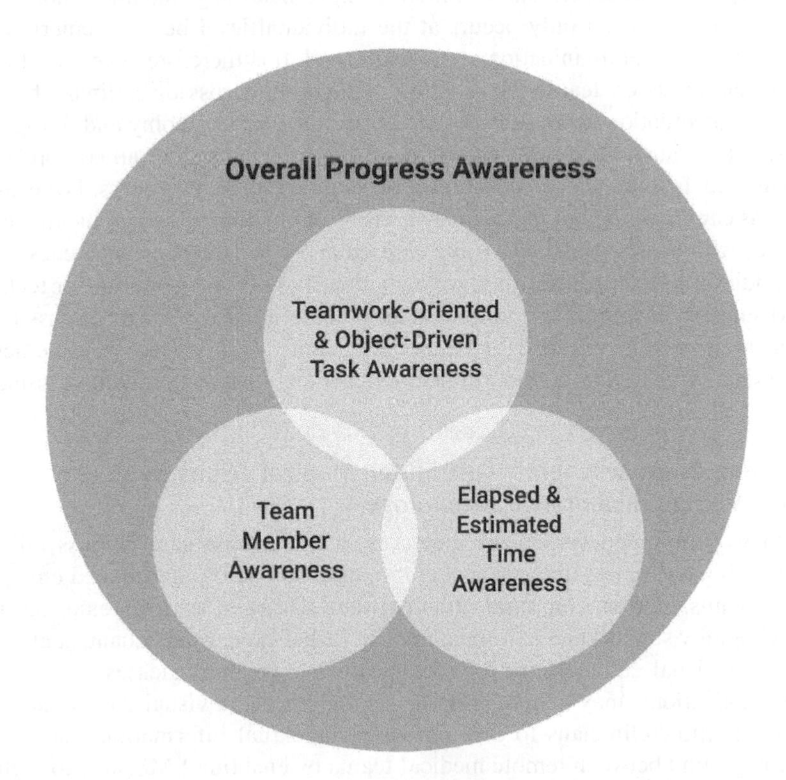

FIGURE 5.3 Overlapping relationships between the facets of awareness.

is multidisciplinary and membership changes often, information and expertise are often highly distributed, and communication across spatial, temporal, and disciplinary boundaries is often fragmented and unsystematic. To overcome these challenges, medical professionals leverage such mechanisms as implicit and explicit coordination, overhearing or situation monitoring, and interpersonal communication to maintain awareness and ensure effective teamwork in dynamic environments. Even so, challenges in awareness and teamwork still persist in the trauma care process, as described in our prior work (Kusunoki et al., 2015; Kusunoki & Sarcevic, 2015; Zhang et al., 2017; Zhang & Sarcevic, 2018).

It has become clear that failure to maintain situation awareness inevitably leads to inaccurate decisions with potentially harmful patient outcomes. For example, studies in surgical settings revealed that surgical errors are often caused by incomplete understanding of the situation (Fabri & Zayas-Castro, 2008) and misperceptions (Way et al., 2003). In addition, Mishra, Catchpole, Dale, and McCulloch (2008) reported a strong correlation between surgeons' situation awareness and the medical outcome. However, to date, training and assessment of cognitive factors such as situation awareness is currently lacking in medical curricula. To align with previous work (Gordon, Baker, Catchpole, Darbyshire, & Schocken, 2015), we want to reiterate the importance of clinicians receiving training in so-called nontechnical skills, such as interpersonal competencies (e.g., teamwork, communication, seeking diverse input and feedback) and cognitive competencies (e.g., situation monitoring). Situation awareness not only occurs at the individual level but also emerges from coordination and communication at the team level. It is therefore necessary to place stronger emphasis on teamwork and communication across disciplinary boundaries as part of situation-awareness-directed training. Our taxonomy and fine-grained analysis of situation awareness can be used to create effective and comprehensive training models that includes different aspects of situation awareness. For example, clinicians can be trained to maintain awareness of ongoing tasks and handle temporal issues with overlapping tasks using elapsed and estimated time awareness.

In addition to medical training, we also identify two opportunities for technologies to support awareness in the trauma care domain. Next we will discuss how to support awareness across distributed medical teams (e.g., between paramedics and ED physicians) as well as within a collocated medical team, i.e., trauma resuscitation team.

Supporting Awareness across Distributed Medical Teams through Visual Collaborative Technology

Current workflow and system architectures in the trauma care process fall short of supporting awareness and efficient communication across distributed emergency medical teams. Current communication between EMS and hospital teams has minimal technology support and is limited to only verbal exchanges (Zhang et al., 2017). The use of visual collaborative technologies, such as smart glasses or augmented reality applications, may help address these challenges. A visual collaborative system would allow clinicians to communicate contextual information and establish common ground between remote medical teams by enabling EMS staff to highlight patient injuries and ED physicians to explain how to treat those injuries (Figure 5.4).

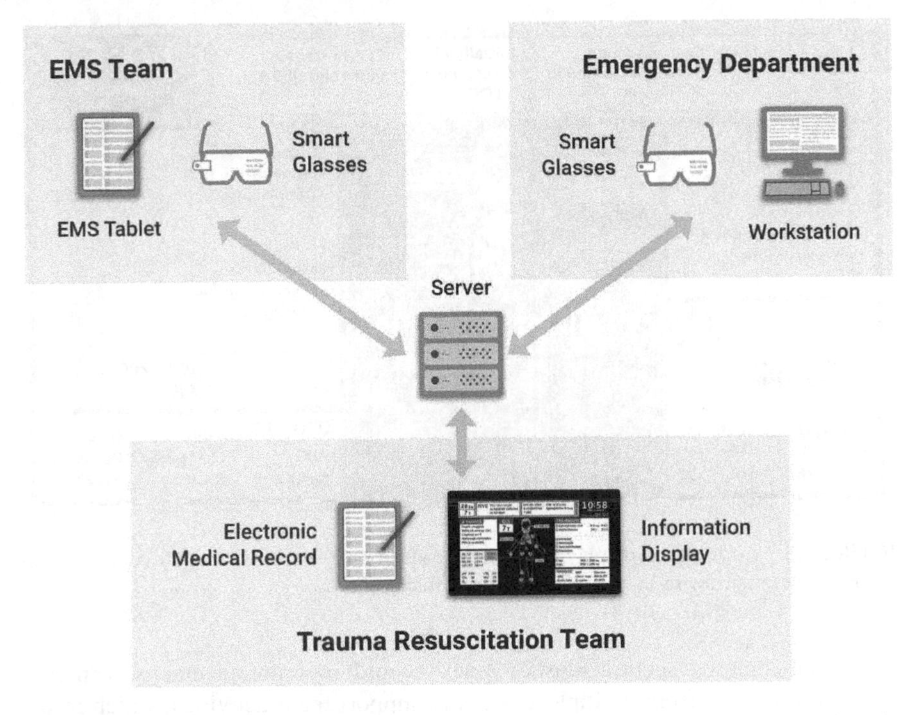

FIGURE 5.4 Envisioned technology solutions to support inter- and intra-team situation awareness.

This could potentially lead to more efficient teamwork, increased awareness, and positive patient outcomes.

Supporting Awareness in Collocated Settings through Information Displays

As an example of collocated and ad hoc teams, the trauma team faces challenges in maintaining awareness similar to those of other fast-paced, multidisciplinary teams, although collective information seeking may help. Moreover, awareness is currently minimally supported by information technology in trauma centers, making information retention and decision making during trauma resuscitations challenging.

Many researchers have examined the use of information displays as potential vehicles to support awareness and collaborative work (e.g., Bardram et al., 2006; Parush et al., 2011; Aronsky, Jones, Lanaghan, & Slovis, 2008; Wears et al., 2007). In our prior work, we designed and developed an information display system for trauma resuscitation teams to construct awareness (Kusunoki et al., 2014). This display has been evaluated and demonstrated that it is feasible to support the construction and management of trauma team members' awareness. As shown in Figure 5.5, the team-centered section (solid lines) would display (1) a list of completed procedures and treatments for teamwork-oriented task awareness and (2) the time of arrival, timer, and timestamps of treatments for elapsed time awareness. The combination of these facets of awareness would then help support overall progress awareness. Moreover,

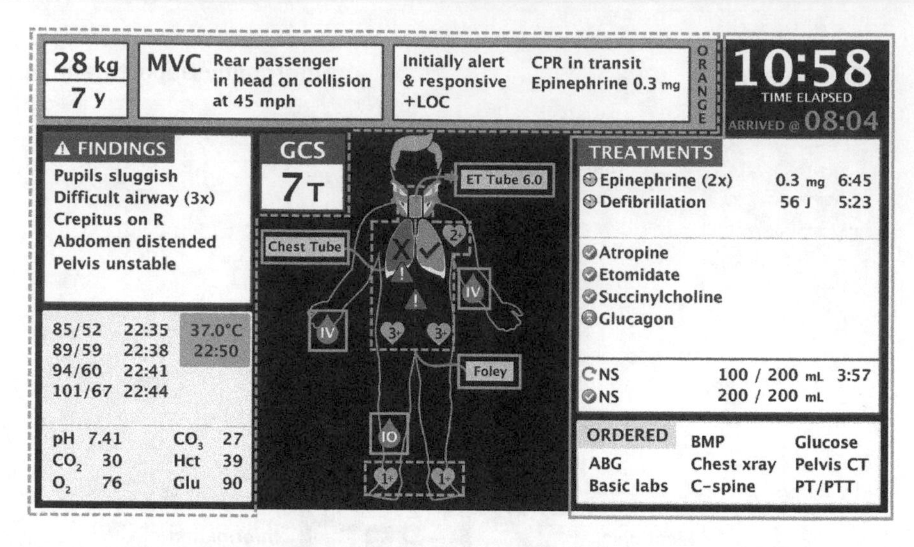

FIGURE 5.5 Information display for trauma resuscitation teams with team-centered (solid lines) and patient-centered information (dashed lines).

the patient-centered section (dashed lines) would incorporate emerging patient-driven information from multiple sources to support team activities, which in turn could support the awareness of trauma resuscitation teams as they coordinate their work in a collocated room.

TOWARD A CONCEPTUAL MODEL OF AWARENESS

Examining the trauma care domain has led us to formulate a conceptual model of awareness—one that draws on several concepts from activity theory (AT)—to understand collaborative work in time-critical medical domains. These concepts include object-orientedness, subject, community, division of labor, tools as mediating artifacts, and rules (Kuutti, 1996). Our conceptual model describes the levels of awareness in relation to the levels of externalization among actors on a team or across teams.

Using Activity Theory as a Theoretical Framework to Understand Teamwork and Situation Awareness In Trauma Care

Previous work by Bardram has used AT to describe the complex work and coordination of activities in the surgical department (Bardram, 1997; Bardram & Doryab, 2011). Similar to Bardram's approach, we propose AT as a possible theoretical framework for understanding trauma care as a sociotechnical system (Table 5.2). AT defines an *activity* situated within a context as the basic unit of analysis. The goal of an activity completed by a *subject* is to transform an *object* into a particular outcome. The *patient* can be considered the object of work, and stabilizing the patient is the goal (object). *Tools* mediate between a *subject* and the object, resulting in the subject

TABLE 5.2
Using Activity Theory to Understand Trauma Care as a Sociotechnical System

	Elements	Description	Examples
	Object	The goal of actions that the object of work that necessitates	Patient stabilization; patient is the object of work
↑↓	**Tools**	Mediating artifacts created by subjects to manage their actions to transform an object	Vital signs monitor, checklist, weight board, medical record
	Subject	A person shaping an object through activity	A clinician
↑↓	**Rules**	Mediates the relationship between subject and community	Stabilization protocol
	Community	A group of subjects sharing the same object	An emergency medical team
↑↓	**Division of labor**	Mediates relationship between object and community	Clinician roles and hierarchy

Note: Based on Kuutti's (1996) description of the structure of an activity. Arrows represent mediating concepts between subjects and the object of work.

transforming the object or managing his or her actions to transform the object. Examples of tools include the medical record, checklist, weight board, vital signs monitor, and the information display proposed in our previous research (Kusunoki et al., 2014). A *community* is a group of subjects who share a common object. There is a *division of labor* among subjects of a community that structures the coordination of activity mediated through *rules*. *Clinicians* are the subjects working together as a community with the shared objective of resuscitating a patient; their labor is distributed by *role*. The main set of rules by which the emergency medical teams (e.g., trauma teams) coordinate their activity is defined by *medical protocols*. Each role is responsible for a general set of *activities* that may or may not be shared with other roles. Activities are also shaped dynamically based on how a patient's status changes over time. Using this framework based on AT, we can further conceptualize awareness in this information space.

Levels of Awareness Information

Similar to the "levels of activity" described by Kuutti (1996), there appear to be *three levels of awareness information* (Figure 5.6): (1) *object level*, at which the object of work is the patient and the objective is patient stabilization; (2) *task level*, involving awareness for completing individual and shared tasks (including task dependencies); and (3) *process level*, looking at the overall state of the trauma care process. Along the horizontal axis is another dimension of awareness ranging from the *subject level* (individual) to the *community level* (group). In collaborative work, awareness information could be viewed as being both *produced* and *consumed* at the subject level

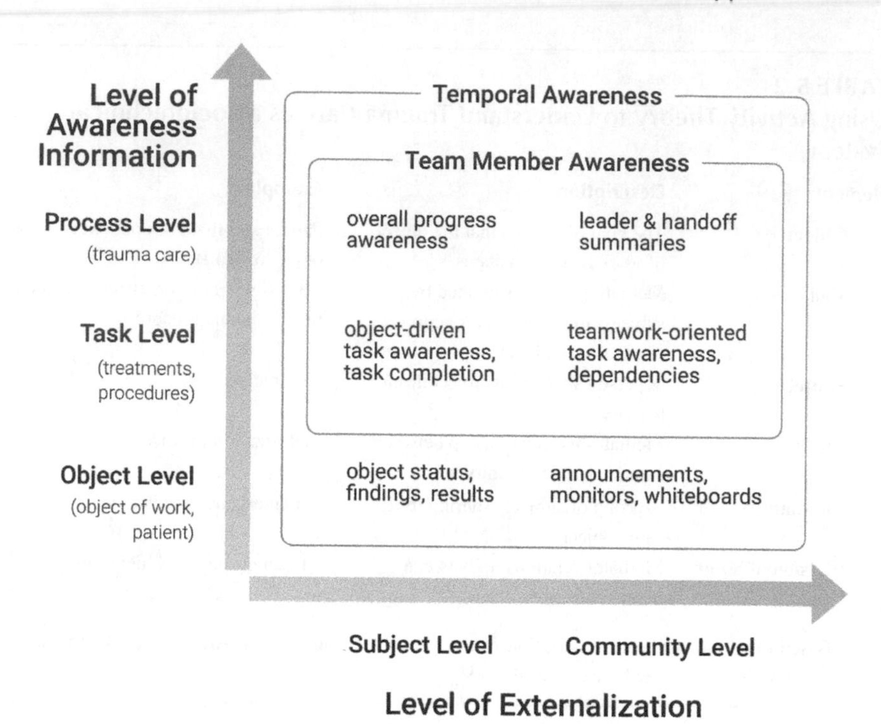

FIGURE 5.6 Conceptual model with spectrums of awareness from the subject to community level.

and then *externalized* either verbally or visually for the community level of awareness. An example of information being *produced* at the subject level of awareness would be the trauma team physician surveyor discovering a large portion of the *object level* information (e.g., patient injuries), which is externalized through verbal communication to the entire trauma team to support community level of awareness. An example of information being consumed at the subject level of awareness would be trauma team leaders consuming information sought through the other roles in order to produce *process level* information for overall awareness (e.g., announcing summaries to the team).

These three levels of awareness information are different from the three-level model of situation awareness proposed by Endsley (1995), namely perception, comprehension, and projection. We conceptualize that in each phase of SA, people need to be aware of, comprehend, and project the status of the object, task, and process. The three levels of awareness information can augment our current understanding of situation awareness in medical teamwork. For example, trauma team members first need to *perceive* the current patient status (object level), treatments performed (task level), and current procedure of the stabilization protocol (process level). They must then *comprehend* all of this information to make sense of what is going on and how things are going. Finally, they make decisions about what to do next and *project* future actions (e.g., who is doing what at which time).

Four Facets of Awareness as Described by the Conceptual Model

The four facets of awareness discussed in the previous section can also be conceptualized within the information space we described previously (Figure 5.6). Awareness of information at the *object level* includes patient status, findings, and results—some of which is externalized through tools visible to the whole team, including the information displays and vital signs monitor. Awareness of object level information is usually necessary to support awareness at the task level. At the *task level*, clinicians need *object-driven task awareness* of information to complete tasks individually (e.g., exams, treatments, procedures) and *teamwork-oriented task awareness* of information such as task dependencies to complete shared tasks. When moving to the *process level*, task level and object level awareness are necessary to support *overall progress awareness* of the resuscitation, which can be at the subject or community level. *Team member awareness* spans across task level and process level awareness because it is concerned with coordinating tasks and how the team is performing overall. *Temporal awareness* can be applied to the whole spectrum because of the time-critical nature of all information being consumed and produced during the resuscitation process. There is a pervasive need for temporal awareness across all dimensions of situation awareness and externalization.

Summary

The proposed conceptual model can be used to understand time-critical and interdisciplinary teamwork. In particular, it can be used to identify which levels of awareness need support. In the trauma care domain, while there are tangible tools for supporting awareness at the process and community levels (i.e., public information display for the team) and at the subject and process levels (i.e., medical record updated by the scribe nurse), there are no apparent tools for supporting awareness at the task level. This lack of awareness support is important to address in the design of new collaborative technologies.

There are some interesting questions to explore in future research to extend this conceptual model of awareness. For example, it is worth exploring the dynamics of situation awareness as it moves between the subject and community levels. How does the intentionality of producing information for collective awareness differ (or not) from the intentionality of consuming awareness information (i.e., selecting or seeking awareness information) for individual use? Studying this phenomenon would involve examining similarities and differences in information consumption and production both *between* and *within* roles on collocated teams, as well as *across* distributed teams, to achieve the different levels of awareness.

CONCLUSION

In this chapter, we have reviewed the awareness research in medical settings. Among the many different facets of awareness that have been proposed and discussed in prior work, we focus on five facets that specifically relate to the characteristics of awareness in medical settings: social, temporal, spatial, activity, and process awareness.

As demonstrated, maintaining awareness in medical settings is a complex and ongoing activity. Several technologies have since been developed to capture, process, and present awareness information.

While these facets of awareness and supporting technologies have been described in detail with regard to collocated, distributed, synchronous, asynchronous, and multidisciplinary contexts, few studies directly addressed the details of awareness in ad hoc, time-critical contexts. We adapt the five facets of awareness to examine what situation awareness means within the context of an ad hoc, time-critical medical setting that features both collocated and distributed teamwork. Further investigation of these micro-level details about situation awareness allows us to not only propose more meaningful mechanisms for supporting the awareness of emergency medical teams, but also formulate a conceptual model to understand awareness and collaborative work of collocated and distributed teams.

REFERENCES

Adams, M. J., Tenney, Y. J., & Pew, R. W. (1995). Situation awareness and the cognitive management of complex systems. *Human Factors*, *37*(1), 85–104.

Aronsky, D., Jones, I., Lanaghan, K., & Slovis, C. M. (2008). Supporting patient care in the emergency department with a computerized whiteboard system. *Journal of the American Medical Informatics Association*, *15*(2), 184–194.

Bardram, J. E. (1997). Plans as situated action: An activity theory approach to workflow systems. In *Proceedings of the Fifth European conference on computer supported cooperative work* (pp. 17–32). Dordrecht: Springer Netherlands.

Bardram, J. E. (2000). Temporal coordination—on time and coordination of collaborative activities at a surgical department. *Computer Supported Cooperative Work*, *9*(2), 157–187.

Bardram, J. E., & Doryab, A. (2011). Activity analysis: Applying activity theory to analyze complex work in hospitals. In *Proceedings of the ACM conference on computer supported cooperative work* (pp. 455–464). New York: ACM.

Bardram, J. E., & Hansen, T. R. (2004). The AWARE architecture: Supporting context-mediated social awareness in mobile cooperation. In *Proceedings of the ACM conference on computer supported cooperative work* (pp. 192–201). New York: ACM.

Bardram, J. E., & Hansen, T. R. (2010a). Why the plan doesn't hold: A study of situated planning, articulation and coordination work in a surgical ward. In *Proceedings of the ACM conference on computer supported cooperative work* (pp. 331–340). New York: ACM.

Bardram, J. E., & Hansen, T. R. (2010b). Context-based workplace awareness. *Computer Supported Cooperative Work*, *19*(2), 105–138. https://doi.org/10.1007/s10606-010-9110-2

Bardram, J. E., Hansen, T. R., & Soegaard, M. (2006). AwareMedia: A shared interactive display supporting social, temporal, and spatial awareness in surgery. In *Proceedings of the ACM conference on computer supported cooperative work* (pp. 109–118). New York: ACM.

Berndtsson, J., & Normark, M. (1999). The coordinative functions of flight strips: Air traffic control work revisited. In *Proceedings of the international ACM conference on supporting group work* (pp. 101–110). New York: ACM.

Bossen, C. (2002). The parameters of common information spaces: The heterogeneity of cooperative work at a hospital ward. In *Proceedings of the ACM conference on computer supported cooperative work* (pp. 176–185). New York: ACM.

Burd, R. S., & Elliot, S. (2011). Evaluation, stabilization and initial management after multiple trauma. In B. P. Fuhrman & J. J. Zimmerman (Eds.), *Pediatric critical care*. St. Louis, MO: Mosby.

Cabitza, F., Sarini, M., & Simone, C. (2007). Providing awareness through situated process maps: The hospital care case. In *Proceedings of the international ACM conference on supporting group work* (pp. 41–50). New York: ACM.

Cabitza, F., & Simone, C. (2007). ". . . and do it the usual way": Fostering awareness of work conventions in document-mediated collaboration. In *Proceedings of the European conference on computer supported cooperative work* (pp. 119–138). London: Springer.

Cabitza, F., Simone, C., & Zorzato, G. (2009). PRODOC: An electronic patient record to foster process-oriented practices. In *Proceedings of the European conference on computer supported cooperative work* (pp. 85–104). London: Springer.

Carroll, J. M., Neale, D. C., Isenhour, P. L., Rosson, M. B., & McCrickard, D. S. (2003). Notification and awareness: Synchronizing task-oriented collaborative activity. *International Journal of Human-Computer Studies*, *58*(5), 605–632.

Carroll, J. M., Rossen, M. B., Farrooq, U., & Xiao, Y. (2009). Beyong being aware. *Information and Organization*, *19*(3), 162–185.

Convertino, G., Mentis, H. M., Rosson, M. B., Carroll, J. M., Slavkovic, A., & Ganoe, C. H. (2008). Articulating common ground in cooperative work: Content and process. In *Proceedings of the ACM conference on human factors in computing systems* (pp. 1637–1646). New York: ACM.

Dourish, P., & Bellotti, V. (1992). Awareness and coordination in shared workspaces. In *Proceedings of the ACM Conference on computer supported cooperative work* (pp. 107–114). New York: ACM.

Drews, F. A., & Westenskow, D. R. (2006). The right picture is worth a thousand numbers: Data displays in anesthesia. *Human Factors*, *48*(1), 59–71.

Durso, F. T., Hackworth, C. A., Truitt, T. R., Crutchfield, J., Nikolic, D., & Manning, C. A. (1998). Situation awareness as a predictor of performance for en route air traffic controllers. *Air Traffic Control Quarterly*, *6*(1), 1–20.

Endsley, M. R. (1995). Toward a theory of situation awareness in dynamic systems. *Human Factors*, *37*(1), 32–64. https://doi.org/10.1518/001872095779049543

Endsley, M. R. (1999). Situation awareness in aviation systems. In *Handbook of aviation human factors* (pp. 257–276). Boca Raton, FL: CRC Press.

Endsley, M. R., & Jones, W. M. (2001). A model of inter- and intrateam situation awareness: Implications for design, training and measurement. In M. McNeese, E. Salas, & M. Endsley (Eds.), *New trends in cooperative activities: Understanding system dynamics in complex environments* (pp. 46–67). Santa Monica, CA: Human Factors and Ergonomics Society.

Fabri, P. J., & Zayas-Castro, J. L. (2008). Human error, not communication and systems, underlies surgical complications. *Surgery*, *144*(4), 557–563.

Favela, J., & Martinez-Garcia, A. I. (2003). Context-aware mobile communication in hospitals. *Computer*, *9*, 38–46.

France, D. J., Levin, S., Hemphill, R., Chen, K., Rickard, D., Makowski, R., . . . Aronsky, D. J. (2005). Emergency physicians' behaviors and workload in the presence of an electronic whiteboard. *International Journal of Medical Informatics*, *74*(10), 827–837.

Gordon, M., Baker, P., Catchpole, K., Darbyshire, D., & Schocken, D. (2015). Devising a consensus definition and framework for non-technical skills in healthcare to support educational design: A modified Delphi study. *Medical Teacher*, *37*(6), 572–577.

Gross, T. (2013). Supporting effortless coordination: 25 years of awareness research. *Computer Supported Cooperative Work*, *22*(4–6), 425–474.

Gutwin, C., & Greenberg, S. (2002). A descriptive framework of workspace awareness for real-time groupware. *Computer Supported Cooperative Work, 11*(3), 411–446. https://doi.org/10.1023/a:1021271517844

Heath, C., & Luff, P. (1992). Collaboration and control: Crisis management and multimedia technology in London underground line control rooms. *Computer Supported Cooperative Work (CSCW), 1*(1–2), 69–94.

Heath, C., Svensson, M. S., Hindmarsh, J., Luff, P., & Vom Lehn, D. (2002). Configuring awareness. *Computer Supported Cooperative Work, 11*(3–4), 317–347.

Hogg, D. N., Folles, K., Strand-Volden, F., & Torralba, B. (1995). Development of a situation awareness measure to evaluate advanced alarm systems in nuclear power plant control rooms. *Ergonomics, 38*(11), 2394–2413.

Hutchins, E. (1995). *Cognition in the wild*. Cambridge, MA: MIT Press.

Kusunoki, D., & Sarcevic, A. (2015). Designing for temporal awareness: The role of temporality in time-critical medical teamwork. In *Proceedings of the ACM conference on computer supported cooperative work & social computing* (pp. 1465–1476). New York: ACM.

Kusunoki, D., Sarcevic, A., Weibel, N., Marsic, I., Zhang, Z., Tuveson, G., & Burd, R. S. (2014). Balancing design tensions: Iterative display design to support ad hoc and interdisciplinary medical teamwork. In *Proceedings of the ACM conference on human factors in computing systems* (pp. 3777–3786). New York: ACM.

Kusunoki, D., Sarcevic, A., Zhang, Z., & Yala, M. (2015). Sketching awareness: A participatory study to elicit designs for supporting Ad Hoc emergency medical teamwork. *Computer Supported Cooperative Work, 24*(1), 1–38. https://doi.org/10.1007/s10606-014-9210-5

Kuutti, K. (1996). Activity theory as a potential framework for human-computer interaction research. In *Context consciousness: Activity theory human-computer interaction* (p. 1744). Cambridge, MA: The MIT Press.

Lai, F., Spitz, G., & Brzezinski, P. (2006). Gestalt operating room display design for perioperative team situation awareness. *Studies in Health Technology and Informatics, 119*, 282–284.

Lane, J. S., Sandberg, W. S., & Rothman, B. (2012). Development and implementation of an integrated mobile situational awareness iPhone application VigiVU TM at an academic medical center. *International Journal of Computer Assisted Radiology and Surgery, 7*(5), 721–735.

Lee, S., Tang, C., Park, S. Y., & Chen, Y. (2012). Loosely formed patient care teams: Communication challenges and technology design. In *Proceedings of the ACM conference on computer supported cooperative work & social computing* (pp. 867–876). New York: ACM.

Levine, W. C., Meyer, M., Brzezinski, P., Robbins, J., Lai, F., Spitz, G., & Sandberg, W. S. (2005). Usability factors in the organization and display of disparate information sources in the operative environment. In *Proceedings of American Medical Informatics Association Annual Symposium* (p. 1025). Bethesda, MD: American Medical Informatics Association.

Ludwig, S., & Lavelle, J. (2010). Resuscitation-pediatric basic and advanced life support. In *Textbook of pediatric emergency medicine* (6th ed., pp. 1–31). Philadelphia, PA: Wolters Kluwer Health/Lippincott Williams, & Wilkins.

Majchrzak, A., More, P. H., & Faraj, S. (2012). Transcending knowledge differences in cross-functional teams. *Organization Science, 23*(4), 951–970.

Mishra, A., Catchpole, K., Dale, T., & McCulloch, P. (2008). The influence of non-technical performance on technical outcome in laparoscopic cholecystectomy. *Surgical Endoscopy, 22*(1), 68–73.

Mitchell, S., Spiteri, M. D., Bates, J., & Coulouris, G. (2000). Context-aware multimedia computing in the intelligent hospital. In *Proceedings of the 9th ACM SIGOPS European workshop: Beyond the PC: New challenges for the operating system* (pp. 13–18). New York: ACM.

Mouloua, M., Gilson, R., Kring, J., & Hancock, P. (2001). Workload, situation awareness, and teaming issues for UAV/UCAV operations. In *Proceedings of the Human Factors and Ergonomics Society annual meeting* (Vol. 45, No. 2, pp. 162–165). Los Angeles, CA: SAGE.

Parush, A., Kramer, C., Foster-Hunt, T., Momtahan, K., Hunter, A., & Sohmer, B. (2011). Communication and team situation awareness in the OR: Implications for augmentative information display. *Journal of Biomedical Informatics*, *44*(3), 477–485. https://doi.org/10.1016/j.jbi.2010.04.002

Prinz, W. (1999). NESSIE: An awareness environment for cooperative settings. In *Proceedings of the European conference on computer supported cooperative work* (pp. 391–410). Dordrecht: Springer.

Reddy, M. C., & Dourish, P. (2002). A finger on the pulse: Temporal rhythms and information seeking in medical work. In *Proceedings of the ACM conference on computer supported cooperative work* (pp. 344–353). New York: ACM.

Reddy, M. C., Dourish, P., & Pratt, W. (2006). Temporality in medical work: Time also matters. *Computer Supported Cooperative Work*, *15*(1), 29–53.

Reddy, M. C., Paul, S. A., Abraham, J., McNeese, M., DeFlitch, C., & Yen, J. (2009). Challenges to effective crisis management: Using information and communication technologies to coordinate emergency medical services and emergency department teams. *International Journal of Medical Informatics*, *78*(4), 259–269.

Sarcevic, A., Marsic, I., Waterhouse, L. J., Stockwell, D. C., & Burd, R. S. (2011). Leadership structures in emergency care settings: A study of two trauma centers. *International Journal of Medical Informatics*, *80*(4), 227–238. https://doi.org/10.1016/j.ijmedinf.2011.01.004

Schmidt, K., & Bannon, L. (1992). Taking CSCW seriously. *Computer Supported Cooperative Work (CSCW)*, *1*(1–2), 7–40.

Scupelli, P. G., Xiao, Y., Fussell, S. R., Kiesler, S., & Gross, M. D. (2010). Supporting coordination in surgical suites: Physical aspects of common information spaces. In *Proceedings of the ACM conference on human factors in computing systems* (pp. 1777–1786). New York: ACM.

Spanjersberg, W., Bergs, E., Mushkudiani, N., Klimek, M., & Schipper, I. B. (2009). Protocol compliance and time management in blunt trauma resuscitation. *Emergency Medicine Journal*, *26*(1), 23–27.

Strater, L. D., Cuevas, H. M., Connors, E. S., Ungvarsky, D. M., & Endsley, M. R. (2008). Situation awareness and collaborative tool usage in ad hoc command and control teams. In *Proceedings of the Human Factors and Ergonomics Society annual meeting* (pp. 468–472). Los Angeles, CA: SAGE Publications.

Strater, L. D., Scielzo, S., Lenox-Tinsley, M., Bolstad, C. A., Cuevas, H. M., Ungvarsky, D. M., Endsley, M. R. (2009). Tools to support ad hoc teams. In P. McDermott & L. Allender (Eds.), *Advanced decision architectures for the warfighter: Foundation and technology* (pp. 359–377). Adelphi, MD: Partners of the Army Research Laboratory.

Strauss, A. (1988). The articulation of project work: An organizational process. *Sociological Quarterly*, *29*(2), 163–178.

Suchman, L. (1997). Centers of coordination: A case and some themes. In *Discourse, tools and reasoning* (pp. 41–62). Berlin and Heidelberg: Springer.

Svensson, M. S., Heath, C., & Luff, P. (2007). Instrumental action: The timely exchange of implements during surgical operations. In *Proceedings of the European conference on computer supported cooperative work* (pp. 41–60). London: Springer.

Tellioğlu, H., & Wagner, I. (2001). Work practices surrounding PACS: The politics of space in hospitals. *Computer Supported Cooperative Work, 10*(2), 163–188.

Wallace, J. R., Scott, S. D., Lai, E., & Jajalla, D. (2011). Investigating the role of a large, shared display in multi-display environments. *Computer Supported Cooperative Work, 20*(6), 529.

Way, L. W., Stewart, L., Gantert, W., Liu, K., Lee, C. M., Whang, K., & Hunter, J. G. (2003). Causes and prevention of laparoscopic bile duct injuries: Analysis of 252 cases from a human factors and cognitive psychology perspective. *Annals of Surgery, 237*(4), 460–469.

Wears, R. L., Perry, S. J., Wilson, S., Galliers, J., & Fone, J. (2007). Emergency department status boards: User-evolved artefacts for inter-and intra-group coordination. *Cognition, Technology & Work, 9*(3), 163–170.

Wilson, S., Galliers, J., & Fone, J. (2006). Not all sharing is equal: The impact of a large display on small group collaborative work. In *Proceedings of the ACM conference on computer supported cooperative work* (pp. 25–28). New York: ACM.

Zerubavel, E. (1979). *Patterns of time in hospital life: A sociological perspective.* Chicago, IL: University of Chicago Press.

Zhang, Z., & Sarcevic, A. (2015). Constructing awareness through speech, gesture, gaze and movement during a time-critical medical task. In *Proceedings of the European conference on computer supported cooperative work* (pp. 163–182). Cham: Springer.

Zhang, Z., & Sarcevic, A. (2018, November). Coordination mechanisms for self-organized work in an emergency communication center. *PACM on Human-Computer Interaction, 2*(CSCW), article 199, 21 pages.

Zhang, Z., Sarcevic, A., & Bossen, C. (2017). Constructing common information spaces across distributed emergency medical teams. In *Proceedings of the ACM conference on computer supported cooperative work* (pp. 934–947). New York: ACM.

6 Team Dynamics of Cybersecurity

Challenges and Opportunities for Team Cognition

Vincent Mancuso and Sarah McGuire

CONTENTS

INTRODUCTION

In December of 2015, consumers of several Ukrainian power companies, totaling around 230,000 people, were left without electricity for up to six hours, following a cyber-attack that switched off numerous substations throughout the country (Zetter, 2016). This outcome was part of a carefully crafted campaign, involving numerous acts of spearphishing, taking command and control of supervisory control and data acquisition (SCADA) systems, destruction of computing infrastructure and data, and a denial of service attack to create mass confusion and uncertainty. In a single coordinated campaign, adversaries not only impacted computing network systems, but also caused kinetic effects across a nation's power grid. In addition, this attack demonstrated the potential for cyber-attacks to impact human behavior, an element of cyber-attacks that has become an important reality to the American public through the topic of election interference.

While not the first instance of such an attack, this incident has become a seminal example of the breadth of impact and interdependencies of cyber on other national assets. What was once considered to be part of the information domain of warfare has since evolved to an interconnected system which includes control of computing systems, critical infrastructure, resources, and human actors. While the vectors of many modern cyber-attacks may be through computing systems and networks, the effects often materialize in other domains. With this realization, cybersecurity and operations have become a pivotal discussion within our national security organizations, from both an operational and research perspective.

In response to this, over the last decade, researchers and practitioners across the government, industrial base, and academia have become invested in establishing a secure and resilient cyberspace. With new threats emerging on a daily basis, significant investments into research and development for new cybersecurity capabilities are being made. Currently much of the research in cyber is grounded in the computer science, mathematics, and engineering fields, with a strong focus on developing algorithms, architectures, and visualizations to help improve detection of malicious activity on a network. The result of this work has led to numerous competencies and capabilities to aid in the defense of our cyber infrastructure and improve our national security posture. While this work continues to have utility, the human element of cyber is equally as critical an element of cybersecurity.

As there continue to be strides made in developing technological competencies to support cybersecurity, our understanding and support to the human-in-the-loop in such environments is still immature. From a human-centered viewpoint, cyber is a cognitively demanding profession. Cyber analysts and defenders rely primarily on their own cognitive resources to interpret observed network activity and translate it into tangible information surrounding the potential of an attack and options for remediation. The majority of this synthesis activity takes place internally in the human mind, placing a significant cognitive burden on them: "the cognitive skill involved in detecting relationships is so critical that any procedures or aids that can expedite or enhance it would improve the analysis process" (D'Amico et al., 2005). In the event of an attack or threat, cyber defenders must identify the adversaries and their capabilities, identify the intended victims, and formulate a hypothesis of the intent of the attack (Caltagirone et al., 2013). During their everyday work, cyber defenders must maintain a mental model of "normal network behavior" despite a constantly shifting landscape: critical networked infrastructure may change without warning, tasking may be rerouted, and agile adversaries are constantly adapting their strategy to elude detection.

If these tasks were not difficult enough, the dynamic and intangible nature of the cyber environment, often referred to as cyberspace, creates a complex and challenging cognitive domain. Cyberspace consists of numerous unique and overlapping networks, nodes within those networks (i.e. any logical device with a network address or analogous identifier, including computing resources that produce, consume, or route data packets, constitutes a node), associated system data (i.e. routing tables, access control lists, etc.), and applications. Across each of these, cyberspace is defined in three layers: *physical*, *logical*, and *cyber-persona*.

- The *physical* layer of cyberspace comprises the physical equipment that transports data and the nodes or systems that process, store, or prepare data for transport. These entities are geographically located in the physical world and are subject to kinetic effects.
- The *logical* layer consists of information and computing assets and their relationships. Together the physical and logical layers can be described as the "terrain" of the cyber domain.
- The *cyber-persona* layer is the digital representation of the identity, authorization, and activities of individual entities (human or computer controlled) in cyberspace, including their behavior in the logical layer and applicable permissions and credentials. This layer is most easily understood as the entities that make use of the network.

Increasing the complexity, cyberspace is a dual role environment in which effects can be achieved, and is an environment for sharing information and synchronization of organizational resources. As organizations often rely on cyberspace resources for their everyday operations, its protection has significant importance to achieving organizational goals.

While these definitions provide structure, they still suffer from complexities that have defining features that separate themselves from each other, making them distinct entities. Unlike the traditional domains of effects, land, sea, and air, cyberspace exists on a digital plane rather than a physical one (Fairfax, Laing, & Vickers, 2015). Cyberspace is an intangible environment, in which the tactical space where operations occur can align with one or many of the traditional effect domains. Where other types of security have physical anchors in the environment, cyberspace exists as a combination of hosts, switches, routers, and entities which are not attached to any physical locations such as the logical, cyber-persona, and supervisory planes of cyberspace (Fanelli & Conti, 2012).

Without physical anchors, analysts are free to create their own mental schema of the environment, based on their own assumptions, biases, and tasking. Additionally, this can result in analysts developing and pivoting between multiple mental models of the network, depending on their current situation. While there is some benefit to having multiple mental models, there are also numerous drawbacks. With the lack of physical anchors, it can be difficult for analysts and commanders to understand the impact of their actions on the environment.

Teamwork in Cybersecurity

Many of the tasks that exist within cybersecurity, such as monitoring networks, host-based analyses, and incident response, are part of a larger sociotechnical system, dependent on numerous actors across multiple organizations (Cooke et al., 2013; Tyworth, Giacobe, Mancuso, & Dancy, 2012). Effective team collaboration in cybersecurity relies on accurate and shared situational awareness of command priorities, network health and status, and resource allocation and tasking. After completing analysis, cyber defenders must efficiently communicate their findings up the chain of command for a decision on an effective course of action. Research has shown that when engaging in tightly coupled collaborations, versus more loose collaborations, cyber defense analysis teams perform better and demonstrate lower biases in their decision making (Rajivan et al., 2013). In their cognitive task analysis of cyber analysts, D'Amico et al. (2005) extracted six unique roles, that are often distributed across numerous analysts, triage, escalation analysis, correlation analysis, threat analysis, incident response, and forensic analysis. In their model, these activities are distributed across three primary stages, detection, situation assessment, and threat assessment, which map to the tactical and strategic stages of warfare. As these activities are distributed across a sociotechnical process, information sharing is critical to ensure that at each phase, the responsible analyst has the most current and relevant situation awareness information. Tyworth et al. (2012) demonstrated this information sharing is not as simple as to be expected, with analysts with different specializations sharing information haphazardly, without providing necessary information, referred to as "throwing [reports] over the metaphorical wall" between organizations. Whenever these breakdowns in team collaboration occurred, organizational-level cyber performance deteriorated, with analysts losing track of critical data, operating in a degraded decision space, and creating incomplete and ineffective boundary objects, which hindered future collaborations. On the other hand, when teams are able to effectively and efficiently share information, even without explicit communication, they are able to outperform teams with less effective collaborative processes (e.g. Buchler, Rajivan, Marusich, Lightner, & Gonzalez, 2018).

Needless to say, teams play a critical role in cybersecurity (Champion, Rajivan, Cooke, & Jariwala, 2012), however the organizational, geographic, and technological boundaries are a deterrent to collaboration amongst analysts. In their study, Tyworth et al. (2012) found that as a result of poor team composition, lack of awareness across the organization, and boundaries between analysts, there was very little communication and information sharing across functional domains. As the information moves up the organizational hierarchy, managers who are being reported to may not necessarily be cyber experts; thus, the analysts and defenders must extract and abstract information into a decision-quality product from raw data (Staheli et al., 2016).

Inter-organizational teams face additional challenges in accessing key information. Providing a framework that permits the free flow of information between organizations while maintaining organizational security and adhering to information security policy is a challenge to collaborative analysis (Hui et al., 2010). Often incidents require correlation of activity across multiple levels of ownership (or classification in the government), enclaves, and organizations to fully understand the impact or identify trends, creating further barriers to effective collaboration. These coordinative challenges are further compounded by the local needs of the cyber defenders in juxtaposition to the global needs of network operations.

Current research in cyber focuses on issues such as developing algorithms and intelligent systems to enable a secure cyberspace, generally brushing aside issues such as awareness and performance metrics, or even a common definition for "cyber." Research continues to ignore the human, even though originating definitions of cyber situation awareness identified their role as critical (Bass, 2000). Regardless of the efficacy of current and future systems, the human element remains an essential part of the system, thus a new discussion on human-centered cyber cognition is necessary.

Cybersecurity is a largely interdependent social system, in which analysts must share information within and across organizational boundaries, collaborate, and make decisions in order to ensure the security of the network (Staheli et al., 2016). Analysts within cybersecurity are often distributed on several axes. From a functional standpoint, cybersecurity is often distributed across multiple analysts with focus on different expertise domains such as intrusion detection, threat analysis, policy, and operational considerations (Tyworth et al., 2012). Similarly, in many organizations, cyber is often organized by functional areas, with departments focused on administration, intelligence, operations, planning, command and control, and logistics (with command and control and intelligence being mostly unique to military organizations). Within these structures, there are numerous interdependent tasks, in which analysts must collaborate and share critical information with others across organizational boundaries.

These structures diverge from teams considered within team cognition and associated literature, which often assumes tightly coupled collaborative processes and dependencies that require synchronized behaviors. Currently, there is relatively limited available research on human-centered cyber cognition, mostly due to a lack of understanding, availability, and access to the necessary resources to conduct cyber research. Whether it is due to classification, in government, or trade secrets, in industry, it can be difficult, if not impossible, for researchers to gain access to necessary data, environments, and experts to conduct the foundational background research to fully understand the cognitive and collaborative work at play. Additionally, as much of this research can lead to an identification of pain points and deficiencies within an organization, being able to share such information is equally if not more of a challenge.

Many have called for an increase in research in these areas (Boyce et al., 2011; Knott et al., 2013; Mancuso et al., 2014), and there are many areas that are likely to bear fruit concerning situation awareness and performance (Gutzwiller, Fugate, Sawyer, & Hancock, 2015). However, currently human-centered cyber often focuses on the analytical and decision-making processes of a singular analyst, with less attention on collaboration, information sharing, and shared situation awareness in cyber defense. This lack of research in team cybersecurity has created a significant gap, as cyber defenders are required to find ad hoc methods to transition from individual to team decision making, creating numerous opportunities for future research and applications of the vast amount of prior work in team collaboration, cognition, and dynamics.

PURPOSE

With the continued and growing importance of cybersecurity, we must continue to push beyond a techno-centric perspective in both research and practice. While the individual human-in-the-loop is critically important in cybersecurity, we must approach them from a perspective that they are part of a larger sociotechnical system, with interdependencies and interactions between multiple humans, organizations, and technologies (Endsley & Connors, 2012; Fink, North, Endert, & Rose, 2009; Tyworth et al., 2012).

The goal of this paper is to focus on team dynamics in cybersecurity and reflect on the major issues in such environments, describe how team cognition research can be applied, and then discuss both research and practical implications. Specifically, we will focus on four main issues of team cognition in cyber: team situation awareness, role definitions and staffing, team hand-offs, and team performance measures. Within each of these topics, we will provide an overview of how the challenge manifests itself within cybersecurity, a discussion on how research in team cognition and dynamics may apply, and potential directions for both researchers and practitioners. Building on these areas, we will discuss emerging considerations such as human-machine teaming, which may have implications for how cyber operations are conducted in the near and distance futures.

TEAM COGNITION IN CYBER: CHALLENGES

TEAM CYBER SITUATION AWARENESS

Overview

While cyber is a relatively immature research area for both the human and computational sciences, cyber situation awareness (SA) has emerged as a persistent thread in terms of understanding analyst decision making, algorithm development, and the design of novel visualizations. Of all the human-centered issues, cyber SA has received by far the most attention, with numerous special journal issues, conferences, and grants being funded in this area.

This is not necessarily surprising—SA is frequently cited as a critical contributor to individual and team performance in complex and dynamic environments. Originally coined by former chief scientist of the United States Air Force Mica R. Endsley (1995) during her research on the cognitive process of pilots, the most ubiquitous definition of SA describes it as "the perception of elements in the environment with a volume of time and space, the comprehension of their meaning, and projection of their status in the near future." As it has been demonstrated in numerous domains, experts believe that accurate, timely, and comprehensive cyber SA provides critical input for decision making when engaging an elusive, adaptive adversary in a constantly changing operational environment.

Unlike more kinetic domains, cyber SA is inherently complex, due to the invisible nature of the cyber environment, the emergent and adaptive threat environment, and data flow rates with a high noise to signal ratio, which transcend human cognitive ability (Endsley & Connors, 2014). From a collaborative perspective, team cyber SA can be considered to be the distribution of critical SA information across a team or organization based on hierarchical role (i.e. analyst, supervisor, commander) and specialization and functional areas (i.e. intrusion detection, network security, intelligence, etc.), and shared and communicated in combination between human cognitive and collaborative behavior and information systems and technological artifacts (Staheli et al., 2016; Tyworth et al., 2012).

In their interviews with analysts, Gutzwiller, Hunt, and Lange (2016) identify three critical elements of cyber SA: the network, the world, and the team. These have some overlap with three classes of awareness defined in Department of Defense Joint Publication 3–12 (2018): Cyberspace Operations: network/system awareness, mission awareness, and adversary awareness. *Network/system awareness* is comprised of key information about the critical cyber terrain in accomplishing the mission, including the status of servers and end points across the network. *Mission awareness*, in which the team is a component, pertains to information about the current tasking and objectives, and the progress and work of the team/organization towards that goal. Finally, *adversary awareness* involves an up-to-date understanding of the threat environment and interdependencies that exist on external networks, organizations, or adversaries. In order to make informed, effective decisions, cyberspace analysts across organizations must have an accurate understanding of what is occurring within their own network and in the world at large (including the activity of their adversaries), and an interpersonal awareness of the requisite situational awareness possessed by others in order to make informed accurate decisions. While building and maintaining accurate cyber situation awareness at the individual level is a cognitively demanding task (D'Amico, Tesone, & Whitley, 2005), these challenges are multiplied at the team and organizational level. Research has shown that situation awareness information and the decisions that are associated with it are often distributed across the organization structure (Staheli et al., 2016).

Analysts must identify their adversaries, locate and isolate victims, formulate a hypothesis of the intent of the attack, and project remediation opportunities, while coordinating and collaborating with others over geographic, functional, and organizational boundaries (Caltagirone, Pendergast, & Betz, 2013; Tyworth et al., 2012). In order to accomplish this, analysts must build a shared situational awareness model while dealing with information overload and balancing team structure and dynamics and communication and collaborative processes (Champion et al., 2012). This is further confounded with a lack of current accurate information on the network and adversaries. Many of the decisions an individual is responsible for in cyber require a current and accurate map of the cyber assets that they currently support. These maps are often developed manually in a tedious and error-prone process, making their accuracy and currency a continuous question (Goodall, D'Amico, & Kopylec, 2009). Additionally, with adversaries constantly changing their tools, techniques, and practices (TTPs), analysts are often tasked with building SA in an incomplete and adversarial decision space.

If the act of obtaining individual situational awareness is not already challenging enough, the distribution of expertise and specializations across organizational boundaries adds to the complexity. Even if every single analyst in an operations center has accurate cyber SA, if the necessary information is not shared, fused, and enriched across the team, the overarching team and organizational SA will be incomplete, resulting in a negative impact on decision making and performance.

Unlike more tightly coupled collaborative domains, in cyber, rather than explicit communication, analysts implicitly collaborate through data (e.g., data dumps, intrusion sets, etc.) and incident reports (D'Amico & Whitley, 2008). To make effective decisions at the individual level, it is not only important that they possess the requisite knowledge and awareness of the situation themselves, but also have an interpersonal awareness and shared understanding of others. For example, during their observations and data collections at a collegiate cyber defense challenge, Buchler at al. (2018) found that experienced teams shared fewer face-to-face communications, relying more heavily on role specialization, implicit communication, and information sharing via collaborative software.

Even with numerous interdependencies across individuals, the majority of the analytical work occurs at the individual level. To better understand, lessons can be leveraged from the classical definition of cooperative collaboration offered by Dillenbourg, Baker, Blaye, and O'Malley (1995, p. 2) in which work is "accomplished by the division of labor among participants, as an activity where each person is responsible for a portion of the problem solving," as compared to collaborative work, where there is a "mutual engagement of participants in a coordinated effort to solve the problem together."

Team SA is typically approached through a lens of multiple people who recognize that they are within the same collaborative system performing a tightly coupled task with coordinated action. In this case, a team is a social entity made up of multiple individuals working together with high interdependency, shared values, and a common goal (Dyer, 1984). However, contrary to this, cyber work is often completed independently, but still maintains a common goal amongst workers, which in turn creates shared values and desires. This creates a large interdependent system, in which individuals act independently with varying levels of coordination and cooperation to achieve a common goal.

Applications of Team Cognition

While individual SA is an important factor in cyber, the "big picture" or complete organizational cyber SA is distributed across multiple people and organizations (Tyworth, Giacobe, Mancuso, McNeese, & Hall, 2013). When scaled to the team or organization level, SA can be described as "the sharing of a common perspective between two or more individuals regarding current environmental events, their meanings, and their projected future" (Wellens, 1993). Similarly, Endsley (1989) defines team SA as "the degree to which every team member possesses the situation awareness required for his or her responsibilities" (Endsley, 1989). These models of SA suggest that it is a state of human cognition shared across multiple people that consists of a mental model of the current situation and an assessment of the impact of competing actions on the environment. The decision maker utilizes a separate decision-making loop which fuses their internal mental model with external information (from the environment and other individuals' mental models) to decide which action to perform.

Team research has long studied SA in many complex domains, such as air traffic control (e.g. Endsley & Rodgers, 1994; Kaber, Perry, Segall, McClernon, & Prinzel III, 2006; Rodgers, 2017), aviation (e.g. Jones & Endsley, 1996; Taylor, 2017; Wickens, 2002), and emergency response operations (e.g. Harrald & Jefferson, 2007; McNeese et al., 2005; McNeese, Mancuso, McNeese, Endsley, & Forster, 2014), to name a few. In recent years, research has identified team cyber SA as a critical element in improved performance and decision making in cyber (e.g. D'Amico et al., 2005; Gutzwiller et al., 2015, 2016; Tyworth et al., 2012).

At the team level SA can be further understood using mental model theories, where complimentary mental models, made up for taskwork and teamwork, have interactions leading to improved team performance (e.g. Klimoski & Mohammed, 1994; Mohammed & Dumville, 2001). Within the task mental model, team members must have an understanding of the technology and systems they are interacting with, the requisite knowledge to complete the task, an understanding of processes and strategies, and an awareness and understanding of the environment, its condition, and dynamics. The teamwork mental model, on the other hand, pertains to an understanding on how the technology is distributed across the team, on how the team interacts, and the division of roles, responsibilities, and knowledge. This also has overlap with transactive memory theories, in which performance is driven by a consensus of expertise, specialization across the group, and accuracy of knowledge (Austin, 2003). Transactive memory states that in order for a team to achieve a maximum performance, the team must be made up of individuals with deviations in their specialization, a consensus and awareness of who is an expert in what, and the knowledge at the individual level must be accurate.

Team cognition research shows that in order for teams to leverage their own situation awareness and knowledge, they must also possess awareness of team and its collaborative dynamics. To achieve this awareness, individuals must effectively communicate and share information, both of which have been shown to be a predictor of team performance (DeChurch & Mesmer-Magnus, 2010). These collaborative processes can not only improve SA at the team level, but also help reduce errors, manage task complexity, and reduce ambiguity and stress (Salas, Cooke, & Rosen, 2008). Without these underlying processes, individuals are unable to obtain the necessary information to update their own SA mental model, and cannot make a fully informed decision, having a negative impact on the overall team and organizational performance.

In the context of cyber, when the analysts are distributed across functional, geographic, and temporal boundaries, these collaborations and the development of shared awareness often occurs within technologies such as shared wikis, message boards, and collaborative portals. Using these technologies, the humans rely on implicit and shared artifacts to enable their collaborations, known as "boundary objects." In their seminal paper, Star and Griesemer (1989) define boundary objects as "objects which are both plastic enough to adapt to local needs and constraints of the several parties employing them, yet robust enough to maintain a common identity across sites" (p. 393). As applied to cyber, boundary objects serve as the technological medium for fostering collaborations and helping individuals to ground their own actions within the context of the larger collaborative and interdependent system.

Based on previous research on team SA and related constructs, several directions for future work, both research-focused and applied, begin to emerge.

Research Implications

Currently, one of the biggest problems that exists within cyber SA is the processes around and the ability to share relevant cyber SA information across the numerous boundaries that exist across organizations. Cyber work is often distributed across individuals, teams, and organizations, and a complete SA mental model depends on an awareness of the knowledge, information, abilities, and priorities of others. If an analyst needs critical information to complete their task, and does not know where to obtain it, they will not be able to improve their SA, due to a lack of critical information and knowledge, and thus cannot make fully informed decisions.

The majority of the discussions on teamwork in fields such as organizational psychology, human factors, and social psychology assumes teams with tightly collaborative processes across minimal amounts of boundaries (e.g. collocated rather than geographical, without organizational boundaries, an implicit motivation to share information, etc.). However, the cyber environment offers numerous challenges and complexities, such as distributed workforces, organizational boundaries, and the desire to withhold information for the purpose of secrecy/classification levels, the "need-to-know," maintaining competitive advantages, and shielding outsiders from knowledge of their organization's deficiencies and vulnerabilities. In 2016 in the United States, the Federal Bureau of Investigation (FBI) Crime Complaint Center (IC3) listed a total of 350,000 reported cyber crimes and estimated this number represented only 15% of the total that occurred (FBI, 2016). This lack of information sharing within and across organizations has become a major decrement to our ability to maintain team cyber SA and perform cyber operations with relevant and timely information.

In such a distributed team environment, the question of how teams can and should share information for the purpose of situational awareness development is a critical question. With an existing deluge of data pouring in that cyber defenders are responsible for, providing too much external information (from other individuals) could result in information overload and have a negative impact on performance. On the other hand, too little information could result in poor decision making and continued persistent threats across the network.

When cyber SA is discussed there is a general understanding that the overarching categories revolve around awareness of the network, the adversary, and the current goals of the organization. While effective in communicating the cyber

SA at a general level, when applied to cybersecurity, these categories may be too broad. There is a clear research need for a better understanding of cyber SA across an organizational hierarchy. While there has been some work to better understand cyber SA (e.g. Tyworth et al., 2012, 2013), future work should focus on better understanding how cyber SA is distributed across the team and larger organization. During their interviews with cyber analysts and decision makers, Staheli et al. (2016) showed that across an organizational hierarchy, the types of decisions and information vary in term of scope and impact and better understanding the information needed to make those decisions has enormous impact. Such a focus could combine methods such as cognitive walkthroughs, contextual inquiry, and cognitive task analysis to social network and job analyses to extract what the critical cyber SA for each individual is and then chart the dependencies as they go across an organization.

As this understanding of cyber SA is formed, research can pivot to studying how assisted information sharing technologies can help cyber operators share and locate critical information to form cyber SA. In their development of the CARINA system, Staheli et al. (2016) applied a user-centered design approach to the development of a system that could help improve organizational information sharing. Building on this approach, researchers can leverage emerging AI technologies to help deliver up-to-the-minute information and break down organizational boundaries and hidden knowledge profiles that may emerge between cyber operators and organizations.

Table 6.1 provides a summarization of the exemplar challenges and associated research directions for team cyber SA in cybersecurity, as well as relevant citations from both cyber and team cognition literature.

TABLE 6.1
Summary of Challenges and Research Directions for Team Cyber SA

Challenge	Research Directions	Relevant Citations
Critical information sharing across geographic, functional, and organizational boundaries	• Cross-boundary mental model formation and alignment • Effective information sharing methodologies	Caltagirone et al. (2013) Caltagirone, Pedergast & Betz, 2013 Staheli et al. (2016) Tyworth et al. (2012)
Lack of explicit collaboration, rather relying on data sharing, collaborative systems, and incident reports to share information	• Develop new and understand existing boundary objects for cyber operations	D'Amico and Whitley (2008) Dillenbourg et al. (1995) Dyer (1984) Star and Griesemer (1989)
Lack of complete cyber SA within and across organizations	• Application of team cognitive models, such as transactive memory and team mental models, to cyber SA theory	Austin (2003) D'Amico et al. (2005) Gutzwiller et al. (2015) Klimoski and Mohammed (1994) Mohammed and Dumville (2001) Tyworth et al. (2012)

Practical Implications

Once the exact cyber SA needs of both the individual, team and organization, and their dependencies and interactions are better understood, we can begin to discuss improving the effectiveness of the teamwork. To understand and improve how collaboration occurs within the system, research must be able to identify and extract key boundary objects from the environment. Using this knowledge, researchers must focus on specific behavioral factors that impact information sharing and then prototype and develop information sharing applications and test them in both laboratory and field-based applications. Using a Living Laboratory approach (McNeese, Mancuso, McNeese, Endsley, & Forster, 2013), researchers can first build a baseline for the current state of information sharing within a team or organization and then leverage the knowledge on what their cyber SA needs are to develop reconfigurable prototypes and deploy them in synthetic task environments for iteration. Eventually they can develop operational prototypes to understand how they could impact and improve cyber SA at the individual, team, and organizational levels. This work can help lead to better tooling, both for individual and team-based cyber analysis.

Role Definitions and Staffing

Overview

Teams within cybersecurity can be composed of multiple individuals performing similar tasks, such as a team comprised of all host analysts. However, even within these seemingly homogenous teams, there can be a wide range of specializations with individuals having varying levels of expertise across the wide range of tools and software used within cybersecurity. Teams can also be comprised of multiple functional roles including forensic, threat, and intel analysts (D'Amico & Whitley, 2008; Zimmerman, 2014). Team composition, which is the balance of not only roles but individual characteristics including education, training, and experience, is an important factor in performance. In cybersecurity, however, there is a growing shortage of trained workers to fill positions. This shortage is causing ambiguity in roles and responsibilities as analysts have to complete tasks that should span multiple positions and it is causing increased workload and occupational stress. According to the 2017 Global Information Security Workforce Study, there will be a global shortage of 1.8 million cybersecurity workers by 2022 (ISC^2). A report by Herjavec Group (2017) predicts an even larger shortage, anticipating 3.5 million unfilled jobs by 2021. This gap between supply and demand has increased by 20% compared to a previous forecast made in 2015. The primary reason for the anticipated shortfall is a lack of individuals with the right training to fill positions. In a survey of cybersecurity and IT professionals (Oltsik, 2017), more than 50% stated that staff size in cybersecurity analytics and operations is currently inappropriate. Specific shortages are reported in the areas of proactive threat hunting, prioritizing and investigating alerts, and computer forensics. To fill the demand, it is estimated that one-third of those currently being hired in cybersecurity come from non-cyber, information technology, and engineering backgrounds.

Those without a cybersecurity background are learning on the job. While they are filling a need for more staff, they may not necessarily fit into a specific job role, which can lead to ambiguity in defining the roles and responsibilities of the current workforce. Staffing levels and role ambiguity are elements that could lead to occupational stress. Chappelle et al. (2013) conducted a survey of cyber operators, both active duty military and civilian contractors, and found that insufficient manning, inadequate tools, and keeping up with the rapidly changing threat environment and new technologies are contributing factors to workplace stress. In a survey of cybersecurity specialists at the National Security Agency, Paul and Dykstra (2017), found that both fatigue and frustration increased significantly across an operation. Increased stress can negatively affect both individual and team performance. An increase in staffing and the development of advanced human-machine teaming capabilities that can automate repetitive tasks are needed to tackle this growing challenge of staffing and workload (McCallam, Frazier, & Savold, 2017).

In 2010, the National Initiative for Cybersecurity Education was created to improve cybersecurity awareness education and workforce development (Newhouse, Keith, Scribner, & Witte, 2017; Paulsen et al., 2012). As part of this initiative the NICE Cybersecurity Workforce Framework was developed, which is comprised of three components to describe cybersecurity work roles. There are seven high-level groupings of cybersecurity functions. This includes the categories of Analyze, Collect and Operate, Investigate, Operate and Maintain, Oversee and Govern, Protect and Defend, and Securely Provision. Within the seven groups are a total of 33 specialty areas of cybersecurity work including areas such as Digital Forensics, Cyber Operations, and Vulnerability Management. Specific work roles are then broken down further into 52 categories that provide specific knowledge, skills, and abilities associated with each. Having this type of framework can aid in the development of training and education programs as well as individual job descriptions. However, the framework does not provide guidance on what the diversity of skillsets should be when developing a cybersecurity operations center, forensic analysis team, threat assessment team, etc. within an organization. Therefore, the diversity of individual skillsets that are required for completing certain tasks still requires further research.

Applications of Team Cognition

Given the shortage in cybersecurity workforce, it is essential that organizations are able to retain their staff. Role ambiguity is the lack of understanding in the tasks that need to be completed as part of one's job. Ambiguity about the expectations of a role tends to lead to job dissatisfaction (Kahn, Wolfe, Quinn, Snoek, & Rosenthal, 1964). Abramis (1994) conducted a meta-analysis of studies that examined role ambiguity and found a negative correlation of moderate effect between job satisfaction and role ambiguity. For job performance, analysis was performed separately for studies in which performance was assessed independently, such as supervisor ratings, and those in which performance was self-assessed. A negative correlation was found but it was weak and predominately for self-assessed performance only, suggesting that ambiguity may affect subjective evaluations of one's job but its effect on objective evaluations of performance needs further investigation (e.g. Driskell & Salas,

1991). Beauchamp, Bray, Eys, and Carron (2002) examined role ambiguity within the defensive and offensive responsibilities of athletes and found that ambiguity concerning the scope of responsibilities was the primary predictor of efficacy and performance. They also found that the relationship between ambiguity and performance was reduced when controlling for efficacy as a potential mediator. Similarly, in a survey of employees in a private sector company, Manas et al. (2018) found that role ambiguity negatively affected engagement at work. Decrease in engagement can lead to decreased individual performance, which can affect the team. Occupational stress due to ambiguity in roles and responsibilities can also affect team performance due to a restriction in attentional focus (Driskell, Salas, & Johnston, 1999). When under stress, an individual may experience information overload. This overload can result in an individual focusing only on their own task, neglecting group tasks and social cues that can then result in degraded team performance.

High team performance is dependent on the skillsets of each individual that makes up the team, which includes both functional and team skillsets. Belbin (1993) has defined nine different team roles including plant (creative, source of inspiration), resource investigator (discoverer), coordinator (maintains objectives, protocol), shaper (focused on achieving results), monitor/evaluator (analyzes ideas, decides after discussion), teamworker (holds team together), implementer (organizer), completer/finisher (ensures standards are maintained), and specialist. For these team skillsets there is inconclusive research on whether there is a relationship between the diversity of skillsets and performance. Van de Water, Ahaus, and Rozier (2008) found that there was no difference in performance between balanced and unbalanced teams according to Belbin's categories. However, they stated that the lack of finding could be due to unclear definitions on what a balanced team is. Similarly, Batenburg, van Walbeek, and der Mauer (2013) did not find that diversity of roles corresponded to higher performance. However, these studies were focused on teams' skills and not on the balance of functional knowledge within the teams.

Stewart (2006) conducted a meta-analysis of studies that examined the design or construct of teams and team performance. Heterogeneity was examined in the included studies for multiple characteristics including demographic, personality, and background characteristics including education and revealed little effect on team performance. Aggregated characteristics of a team though including expertise were found to have a positive correlation with performance.

Constructing a team by balancing the skillsets across team members is just one of four team composition models described by Mathieu, Tannebaum, Donsbach, and Alliger (2014). The other three models described include the traditional personnel–position fit model which assumes that teams will be more effective if individuals have the knowledge, skills, and abilities for their respective positions: the personnel model with teamwork considerations, which is based on the idea that team performance is improved through team-related competencies including coordination, communication, leadership, and decision making. This model suggests teams that possess more generic team-related skills will be higher performing than teams with less team skillsets. The fourth model is a contribution model, which assumes that characteristics of a specific individuals may contribute more or less to team performance than others due to their roles and responsibilities.

Which model though leads to highest team performance is inconclusive and more research is needed.

Role ambiguity and team composition can also impair a team's shared mental models. Job dissatisfaction can lead to increased turnover rates which will make it difficult for teams to form shared mental models. If groups have a shared understanding and knowledge of a situation they are more likely to work in a coordinated manner (Klimoski & Mohammed, 1994). Teams that have expertise and experience in common are more likely to process information and act in common ways. Edwards, Anthony Day, Arthur Jr., and Bell (2006) examined the similarity and accuracy of team mental models and how it relates to performance. If a mental model is accurate it is expected to also be similar across a team. However, it is possible that a team may have the same mental model but it is not correct. As expertise increases across the team, it is expected that both accuracy and similarity of the team's mental model will increase. Edwards et al. (2006) examined the performance of three teams with difference expertise levels and found that approximately one-quarter of the variance in team performance was accounted for by the team skillset/expertise composition.

Research Implications

Further research is needed on the skillsets and job roles of cybersecurity teams. Little research has been done on the individual skillsets on a team that will lead to improved performance. Henshel et al. (2016) examined training certifications, education of team members, and performance in cyber exercises and found that having college as the highest level of education and having obtained a cybersecurity training certificate was significantly related to proficiency metrics including the time from inject to the approval for taking action. However, it is not only cyber-specific skills that need to be studied. Ben-Asher and Gonzalez (2015) found in a laboratory study that there was no difference in the ability to detect network attacks between individuals who were considered experts in cybersecurity and novices who had little to no expertise in cybersecurity. Potential explanations for the lack of difference between experts and novices in identifying attacks is that knowledge needs to include both situational knowledge of the environment and domain knowledge. Experts may have a hard time identifying attacks when they are unfamiliar with the network layout and baseline activities within the network. However, it also suggests that further work is needed in cyber on understanding how individuals accumulate information during a decision-making process and what basic cognitive skills are related to successful threat detection.

To fill the current staffing demands, people are entering cybersecurity from other fields. It could be useful to use cognitive aptitude tests to evaluate these candidates for traits that might indicate that they would be successful in the field. Selecting individuals with the right skillsets will lead to improvement in overall performance in a cyber team. A cyber aptitude model has been proposed by Campbell, O'Rouke, and Bunting (2015) and accounts for both critical thinking and job-specific knowledge. For job-specific knowledge, they used the NICE framework discussed in the previous section. For the critical thinking component, they indicated measuring four components including proactive and reactive thinking ability and real-time and deliberate action ability. Further research on which cognitive abilities are most

TABLE 6.2
Summary of Challenges and Research Directions for Role Definitions and Staffing

Challenge	Research Directions	Relevant Citations
Lack of staffing is leading to ambiguity in defined roles and responsibilities within cybersecurity teams	• Understand skillsets and job roles of high-performing cybersecurity teams	Ben-Asher and Gonzalez (2015) Henshel et al. (2016) Oltsik (2017)
Many individuals entering cybersecurity have no prior experience in the field	• Identify cognitive attributes that are associated with high-performing cyber analysts	Campbell et al. (2015) Oltsik (2017)
Varying skillsets and experience levels make it challenging to distribute work	• Development of capabilities for improving distribution of workload across teams	Gersh and Bos (2014)

important for success in the field of cybersecurity, such as pattern recognition, vigilance, and spatial visualization, is needed. Tests for these abilities should be identified or developed and evaluation of performance of individuals determined to have high aptitude based on cognitive skills is needed.

Once a team is formed, it is important to adequately allocate workload across the team. Clearly defining roles and responsibilities can lead to improved work allocation. Further research is needed on the development of tools to improve the triaging of events across analysts so that if one analyst becomes overburdened the workload can be shifted in order to improve overall performance. Novices working in cybersecurity are biased to thinking that most alerts are real and need investigation, while more experienced analysts think about alert criteria, probability of an event, and thresholds for events to trigger when making a judgment (Gersh & Bos, 2014). Novices may spend significant amounts of time investigating alerts that are false alarms, increasing the workload of more senior analysts. Development of tools, particularly larger capabilities used for intrusion detection and prevention, that can help guide and provide assistance to novices as they are learning on the job can aid in the distribution of workload across operators with a range of experience levels.

Table 6.2 provides a summarization of the exemplar challenges and associated research directions for role definitions and staffing in cybersecurity, as well as relevant citations from both cyber and team cognition literature.

Practical Implications

The tactics, techniques, and procedures (TTPs) used by adversaries continues to rapidly change, and with that the diversity of skills required of cyber operators. Teams need to have a heterogeneous set of task and teamwork skills and need to work together in a collaborative way to quickly and efficiently identify and mitigate threats. While frameworks are being developed defining the knowledge skills and abilities for specific job roles, further work on the development of guidance is needed for the skillsets that should make up specific teams within cybersecurity. It is estimated that for every 5,000 alerts, almost 12% are legitimate events that are never investigated or remediated due to lack of staffing (Cisco, 2017). To increase the size of the workforce in cybersecurity, both task-specific knowledge and relevant cognitive skills that are related to high performance should be considered.

Hand-Offs and Cross-Site Coordination

Overview

Hand-offs are the transition of information and responsibility from one individual to another. In cybersecurity, an alert could involve coordination and collaboration across analysts requiring multiple hand-offs within an investigation. These hand-offs may occur within a team or across the organization. Information on an incident may be submitted to an IT or security help desk by a user, with information on time, location, and sequence of events being transferred to an appropriate analyst for investigation. During an investigation there may also be hand-offs among analysts who are jointly working on different components or aspects of the same incident. Depending on the nature of the alert, multiple skillsets may be needed with for example network, host, or forensic analysts jointly working together. Key information needs to be shared in order to effectively complete the investigations, with frequent hand-offs occurring based on evidence found.

Hand-offs of information may also occur across shifts. In some organizations, security operations centers (SOCs) operate 24/7. During shift changes hand-offs of investigations occur from one analyst to another so that they continue to be handled efficiently without delay. It is important that during the hand-off information is accurately provided so that there is not duplication of work that has been already conducted. Daily update briefings may also occur within an organization, in which information on key investigations is shared across entire teams. Such briefings provide opportunities for all individuals including leadership to have shared situational awareness of key events and for coordination across the team. Larger organizations may be geographically dispersed with coordination and information sharing needing to occur across sites. Different sites may have their own workflows and procedures leading to challenges in coordination.

The hand-off of information is not only necessary within an organization for handling an incident—cyber intelligence needs to be shared across organizations. There are many tools and frameworks such as MITRE's Structured Threat Information Expression (STIX), Common Vulnerability Reporting Frame (CVRF), etc. (see also Asgarli & Burger, 2016) that have been and are being developed. These frameworks do not necessarily use the same conventions, which impairs the ability to share information and improve overall performance across multi-organization teams including government agencies, universities, and commercial companies.

Every hand-off of information is a potential source of error in an investigation. Miscommunication can lead to events being mishandled or an investigation being delayed. Hand-offs between shifts are common in other fields, particularly in hospitals, where hand-offs of patient care occur daily. It has been found that 80% of severe medical errors are attributed to miscommunications during the hand-offs of patient care (Joint Commission, 2012). Effective communication across teams is therefore essential throughout the investigative process. There are many methods currently in use in cybersecurity to transfer information. Information may be transferred verbally during face-to-face or phone conversations. Information is often shared electronically using chat tools and email, or more formal ticket systems may be used in which key information fields are filled out at the start of the investigation, with updated information added throughout the investigation. To effectively convey information, it is important for team members to know the roles, tasks, and information requirements of the other analysts (Cannon-Bowers & Salas, 1998).

Applications of Team Cognition

Hand-offs of information have been studied often in relation to patient care in the medical field. Keebler et al. (2016) conducted a meta-analysis that examined the use of different hand-off protocols and found that the amount of information provided during a hand-off increased after a formal protocol was implemented, which resulted in improved patient outcomes and overall improved satisfaction with the hospital or organization. However, Jiang et al. (2017) found conflicting results, with an increase in reported discrepancies in patient care when an electronic hand-off tool was implemented. This result suggests that any hand-off procedure needs to be thoroughly tested and training provided to all staff. In addition, process-oriented changes are also needed to improve the overall hand-off procedure of information.

Hand-off protocols serve as prompts, providing the categories of information that should be included. Having a standard protocol can lead to reduced cognitive bias among individuals on what information should be shared. Cybersecurity analysts examine a high volume of alerts every day, many of which are false positives. Einhorn and Hogarth (1978) state that outcomes are coded based on frequencies. Therefore, based on past experiences, some analysts may be prone to not share information on certain alerts as they are confident based on the frequency of past events that it is a false alarm.

The complexity of information sharing is increased when teams are located across sites or when teams only communicate virtually. O'Leary and Cummings (2007) stated that outcomes of geographically dispersed teams are in general negative, with reduced communications and real-time problem solving and increased coordination complexity. However, there are some potential positive outcomes, such as the sharing of non-redundant information. Espinosa, Nan, and Carmel (2015) examined what aspects of geographic dispersion may affect team performance. Participants completed a map creation task and communication was completed through an electronic chat tool. They found that temporal distance did correlate to an increase in the time it took to complete a task and that distance was negatively associated with accuracy and quality. They found that distance had no effect on the conveyance of information, rather it was positively associated with communication frequency and negatively associated with turn-taking, while both frequency and turn-taking had a positive effect on the convergence of information. Therefore, the results suggest that it is not necessarily the distance but the pattern of communication that impacts team performance. This is supported by Marlow, Lacerenza, Paoletti, Burke, and Salas (2018) who found that frequency and quality of communications among teams is positively related to performance, with a stronger relationship for communication quality.

Oshri, van Fenema, and Kotlarsky (2008) mentioned that communication technologies are not enough to prevent breakdowns in coordination/communication between sites. Rather, there needs to be trust and ties between teams to facilitate information sharing between sites. Distributed sites will often have their own processes. They investigated transactive memory for two projects which had globally distributed teams. As discussed previously, transactive memory is a combination of communication between individuals and the memory processes of individuals (Wegner Erber, & Raymond, 1991; Wegner, 1995) and is tightly linked with team performance, as individuals know who to refer to for specific information. Without co-location or periodic training, however, it may be hard to develop a transitive memory system. Oshri et al. (2008) found that for globally distributed teams, using standard templates helped with obtaining the appropriate information and then transferring it to the other teams. Hansen (1999) had similar findings, in that for the sharing of complex information strong ties are needed within subunits of an organization. For distributed or multi-team systems it is therefore important to have formal mechanisms defined for meetings, schedules, timelines, and other forms of communication for them to be effective (Shuffler & Carter, 2018).

Information sharing during hand-offs should not only contain factual information but an individual's sensemaking of the situation. Uitdewilligen and Waller (2018) examined information sharing in 12 multidisciplinary crisis teams. As part of their study they examined a model of information sharing that includes fact sharing, interpretation sharing, which is a combination of the individual's expertise and their interpretation of events, and projection sharing, which is the sharing of information regarding anticipated or potential future events. High-performing teams spent more time sharing information before making decisions and they also had more sharing of interpretations, i.e. individuals working together to make sense of the situation, than low-performing teams. Faraj and Xiao (2006) examined coordination within a trauma setting. Based on observations and interviews they defined two types of coordination. The first type of coordination is *expertise coordination*, which is related to managing the knowledge and expertise of individuals. This involves knowing who to coordinate with for crucial knowledge, essentially knowing the expertise of team members. In cyber, this could involve knowing who to coordinate with to conduct host-based analysis, to analyze network packets, etc. The other type of coordination is *dialogic coordination*, which is more contextually based and may be unpredictable. As an investigation unfolds, actions may be less predictable, and it may be necessary for skill-based boundaries to be discarded and everyone to coordinate together to solve the problem.

Research Implications

There are multiple methods that are currently used to hand-off investigations. Hand-offs can occur synchronously with information being communicated verbally in person or over the phone, or virtually using chatrooms. However, the majority are likely occurring asynchronously using tools such as ticketing systems or other collaborative

TABLE 6.3
Summary of Challenges and Research Directions for Hand-Offs and Cross-Site Coordination

Challenge	Research Directions	Relevant Citations
Hand-offs occur often across shifts and across analysts during an investigation	• Development and testing of incident hand-off procedures	Keebler et al. (2016) Jiang et al. (2017)
Teams can be globally distributed, each with their own procedures	• Development of tools to improve the sharing of information and communication for distributed cyber teams	Espinosa et al. (2015) Marlow et al. (2018) O'Leary (2007) Shuffler and Carter (2018)
Fact-only-based information shared across teams could lead to misunderstanding or lead to decisions being made based on prior experience	• Research is needed on the content of information that is shared within hand-offs and within information sharing tools; this includes both fact-based information and sensemaking information	Einhorn and Hogarth (1978) Farah and Xiao (2006)

tools. Studies should be conducted examining how different information sharing tools and their level of synchronicity affect communication patterns and ultimately performance in incident handling. Information or collaborative tools are not only useful for sharing information. They can also provide a workspace where individuals can provide information on their searches, analysis methods, and findings. The utility of a such a tool is based on the opportunistic problem-solving model (Hayes-Roth, Hayes-Roth, Rosenchein & Cammarata, 1979) and the architecture for a similar concept that supports collaborative planning has been developed (DeStefano, Lachevet, & Carozzoni, 2008). This has the benefit of allowing the individual to offload their working memory as they may be working on multiple investigations at the same time and it provides a space for multiple analysts to work together, aiding collaborative sensemaking.

Table 6.3 provides a summarization of the exemplar challenges and associated research directions for hand-offs and cross-site coordination in cybersecurity, as well as relevant citations from both cyber and team cognition literature.

Practical Implications

Having a hand-off protocol could reduce reporting bias. Every individual may not consider information to be of the same priority level which can result in high variability in the information that is provided during hand-offs. Having a set protocol can lead to improved shared mental models across staff, as everyone will have an understanding of what information is considered of high importance. However, there can also be negative consequences to standardizing hand-off protocols. Reporting information in a standard format may require more time, further increasing the workload of analysts who are already dealing with high volumes of daily alerts to investigate. In addition, if the hand-off procedure is too rigid, information that may be unique to a specific investigation may not be shared as it does not fit within the protocol. As new threats continue to emerge, it is essential that any procedure allows for flexibility. There is a need for structure for formal coordination and clearly defining responsibilities, however it needs to be a flexible structure that can handle the rapid response needed for incident response in cybersecurity.

Team Measures

Overview

Performance in cyber, like many other topics, is currently considered to be a technological problem. When a new tool or capability for cybersecurity is deployed, individuals have an in-depth understanding of the performance of the technology, how much memory it uses, how much data it requires, its processing speed, etc., with limited insight into the impact on the human. At best, subjective and outcome-based measures are relied on, but more often than not human performance is simply ignored all together.

Measuring human-centered outcomes in cyber is a major challenge. Similar to safety science, cybersecurity metrics are often grounded in outcome: was the threat removed, was there information spillage, what was the data loss, etc. While these types of measures are important in understanding the overall security posture of an organization, they provide little to no real-time feedback to analysts or their supervisors to help them regulate their behaviors and change their processes or strategies. Currently, analysts have to rely on their own instincts to regulate their behaviors or wait for after-action reviews to provide them with feedback on their performance. However, a problem still arises in that if the adversary continues to go undetected, the analysts and organization may have no idea that they are performing poorly, and may continue down that path until it is too late.

Human-centered cyber metrics have major implications on the design, implementation and analysis of cyber exercises, which are becoming common ways to test and field new capabilities and provide training to cyber analysts (Patriciu & Furtuna, 2009; Sommestad & Hallberg, 2012). Whether exercises are fielded for the purpose of evaluation, competition, or training, the understanding of human processes, behaviors, attitudes, and performance is critical.

For evaluation there is a desire to understand the performance of a team with and without a certain tool. In this case, the measures of performance are critical in better informing the continued development or acquisition of future cyber capabilities for that team. In competitions, there is a desire to measure the skills of individuals and teams so that at the conclusion of an event, an assessor can make a judgment on the efficacy of a team and their ability to complete a task. For training, performance is important in showing an improvement to an individual or team and understanding their processes can shed light on their knowledge, skills, and abilities in conducting their tasks. At the conclusion of a cyber exercise, participants are provided with a "hot-wash" of their performance. Currently, much of this is based on performance outcomes: did they find the attacker, did they document the evidence, etc., and sometimes based on augmented subjective ratings by subject matter experts. However, without clear performance and process metrics, participants are unable to understand how their processes and actions may have impacted these outcomes and have no way in the future to regulate their behaviors to improve performance.

Additionally, from a technology development standpoint, outcome measures provide limited insight into how the capability impacted performance. It is possible that the team came to the correct conclusions while using the tool in an incorrect way, or by chance. By leveraging only outcome measures, a team may be assigned a good or bad score, without full insight into whether or not they achieved those outcomes in a positive manner. This could lead to a team forming a poor metacognitive awareness mental model and not regulating their behaviors in situations where they achieved a high score while using incorrect processes, or over-regulating if they used correct processes but achieved a low score (Dinsmore, Alexander, & Loughlin, 2008).

Applications of Team Cognition

Performance itself is typically an outcome-based measure, thus it is important to not only collect accurate performance metrics, but to identify its antecedents and collect appropriate metrics at each level. Typically, in the literature, when discussing different types of team-level metrics, they look at two other areas in addition to team performance, specifically team attitudes and behaviors (Cohen & Bailey, 1997). Based on varying mediation and moderating effects of the team's attitudes and behaviors, team performance represents the outcomes of team collaboration, that is, how well they performed in a given task. Team attitudes, or perceptions, typically capture a team's collective reaction to the teams functioning and performance. Team behaviors include any interpersonal aspects of their collaboration, including communications, planning, and coordination.

Team Performance

Team performance metrics aim at capturing indicators of a team's efficacy at completing a given task. Measuring team performance is very challenging and the difficulty increases with task complexity. In simple laboratory tasks, measures such as accuracy and reaction time may be sufficient indicators in how well an individual is performing, however it is unknown how they may scale to a more naturalistic environment in which the tasks are not as controlled. Regardless, there is no individual silver bullet, as measures need to be combined with other environmental variables to provide a more tangible performance score. For example, the Human Performance Scoring Model (HPSM; Wellens & Ergener, 1988), which interprets team interactions into measurable actions against simulated events based on their accuracy and reaction time, has been used in numerous team-based environments (e.g. Hellar & McNeese, 2010; Mancuso, Minotra, Giacobe, McNeese, & Tyworth, 2012). When moving to more complex tasks, team performance becomes more nuanced and requires a combination of outcome measures with online task-specific knowledge that has direct ties to team performance (i.e. Baker & Salas, 1997; Cannon-Bowers & Salas, 2001; DeChurch & Mesmer-Magnus, 2010; Mathieu, Maynard, Rapp, & Gilson, 2008). These types of measures require a subject matter expert to extract critical information or procedural knowledge from the task and quiz participants on it, with the assumption that if they possess it, they will be making positive progress in their task. For example, the Situation Awareness Global Assessment Technique (SAGAT; Endsley, 1988) queries a team (or individual) on critical knowledge that represents each level of situation awareness that is critical in performing a task. Similarly, accuracy and similarity grids can be used to elicit indicators on the team's mental model during a task session (Cooke, Salas, Cannon-Bowers, & Stout, 2000).

While any individual measure may not provide a full picture, triangulation across measures can provide a more robust picture of team performance. By picking multiple methods, one does not only capture the team performance outcome (as would be the case with a mission performance metric), but additionally captures aspects of the team processes that lead to that outcome and the team's interpretation of their actions. When combined, one can form a complete understanding of the team's performance, how they accomplished their goal, and their mindset in doing so. Such information can be critical in not only evaluating, but also in training and improving team performance and processes.

Team Attitudes

Team attitudes may be the most commonly captured team measurement, possibly due to their ease of use, as they are most often delivered as a survey instrument. Many of the metrics used to assess team attitudes are grounded in research on individuals, however their importance is often magnified when moving to the team level (Cannon-Bowers & Salas, 2001).

One important team attitude that has been shown to impact performance is team emotional state or affect (e.g., Mathieu et al., 2008; Schwarz, 2000). The Positive and Negative Affectivity Scales (PANAS; Watson, Clark, & Tellegen, 1988), and its expanded version the PANAS-X (Watson & Clark, 1999), are the most popular metric for capturing affect. Perceived stress is another example of another team attitude that has been shown to be relevant in assessing team behavior and performance. A common assessment of stress is the Dundee Stress State Questionnaire (DSSQ; Matthews et al., 2002) and its shortened version the Short Stress State Questionnaire (SSSQ; Helton & Garland, 2006). Both the DSSQ and SSSQ are based on Mathews et al.'s (1999) three factors of stress—worry, distress, and engagement—which have been shown to have a correlation with performance. Another construct which shares commonalities with stress that is also relevant to team tasks is workload. To assess workload researchers often utilize the NASA Task Load Index (NASA-TLX; Hart & Staveland, 1988). More recently, Helton, Funke, and Knott (2014) proposed a modified NASA-TLX that includes an additional six items to account for the collaborative requirements of teamwork. Other tools for assessing workload in individual and team tasks include the Bedford Scale (Roscoe, 1987), the Cooper Harper Scale (Cooper & Harper, 1969), the Subjective Workload Assessment Technique (SWAT; Reid & Nygren, 1988) and the Subjective Workload Dominance Technique (SWORD; Vidulich, Ward, & Schueren, 1991), and the Workload Profile (WP; Tsang & Velazquez, 1996).

Team attitudes can also focus on perceptions of cognitive activity, which is shown to be separable from, but equally as important as actual cognition (e.g. Endsley, Selcon, Hardiman, & Croft, 1998; Rousseau, Tremblay, Banbury, Breton, & Guitouni, 2010). Numerous metrics of perceived situation awareness have been applied at the team level, such as the Situation Awareness Rating Technique (SART;

Taylor, 2017), the Crew Awareness Rating technique (CARS; McGuinness & Foy, 2000), the Mission Awareness Rating Technique (MARS; Matthews & Beal, 2002), the Situational Awareness Rating Scale (SARS; Waag & Houck, 1994), and the Situation Awareness Subjective Workload Dominance Technique (SA-SWORD, Vidulich & Hughes, 1991). Perceived situation awareness has been shown to have interacting and mediating effects with actual situation awareness and team performance, and thus has been identified as a critical measure to capture when considering situation awareness in team tasks (Hamilton, Mancuso, Mohammed, Tesler, & McNeese, 2017).

Finally, a team's perceptions of themselves, sometimes referred to as *team viability* (Mathieu et al., 2008), is an important construct in metacognitive regulation of team behaviors. Such attributes are often considered one of the most important antecedents of effective group work (Beal, Cohen, Burke, & McLendon, 2003). These metrics relate to opinions on how well the team perceives their ability to work together and complete their task. The Group Environment Questionnaire (GEQ; Carron, Widmeyer, & Brawley, 1985) has been the standard in measuring cohesion in groups. The survey accounts for four factors of cohesion, group integration, and attractions in regard to task and social work (Cota, Evans, Dion, Kilik, & Longman, 1995). Other metrics that may be captured to assess the team viability include collective efficacy (Riggs & Knight, 1994), perceived collaboration (Lyons, Funke, Nelson, & Knott, 2011), and trust (MacDonald, Kessel, & Fuller, 1972; Mayer & Davis, 1999) to name a few.

Team Behaviors

While the performance and survey metrics described in the previous section can give insight into team knowledge and performance and the team's perceptions of their interactions, they do not account for their actual behaviors. Methods including observer evaluations and communication analyses can capture important aspects of team behaviors. Such metrics are especially useful in complex situations, such as field exercises (e.g. Matthews & Beal, 2002) or real-world situations (Mazzocco et al., 2009), where surveys may not be administered and the experimenter may not have the control necessary to implement metrics for performance.

There are two ways of capturing team behavior from observable ratings: global rating scales and frequency of behaviors. Global rating scales, also known as behaviorally anchored rating scales (Kendall & Salas, 2004), rely on observers to identify key behaviors linked to performance and subjectively rate their effectiveness at that time. An example of this is the Anti-Air Warfare Team Performance Index (Johnston, Smith-Jentsch, & Cannon-Bowers, 1997), in which a subject matter expert identifies key behaviors that indicate superior, adequate, and poor performance. These require observers to simply rate the frequency of various critical actions, without having to make a subjective judgment on the quality.

In addition to their performance-based and survey form, several metrics can also be calculated based on observer ratings. For example, the Situation Awareness Behaviorally Anchored Rating Scale (SABARS; Matthews & Beal, 2002) is an observer rating technique that has been used to assess live infantry exercises. Other attributes of team interactions that have been measured through subject matter expert evaluations include team transactive memory (Smith, 1999), learning (Edmondson, 1999), team coordination, and cohesion (Brannick et al., 1993). In addition to capturing metrics on team interactions, observations can also be used as a way to give insight into the performance of a team during a task (Brannick, Roach, & Salas, 1993).

In an effort to reduce some of the subjectivity, team communications can also be used to understand a team's cognitive and collaborative behaviors (Salas, Bowers, & Cannon-Bowers, 1995). Researchers have posited that enhanced metrics on team communication are the best indicators of team cognition (Cooke, Gorman, Myers & Duran, 2013). Commonly used metrics such as frequency counts or discourse durations can provide information about the amount of information that is being shared during a team task (Volpe, Cannon-Bowers, Salas, & Spector, 1996). These however do not account for the content of the communication, which would require external coders and may introduce subjectivity. Several researchers often rely on task-based coding schemes and include categories that are specific to the given task (e.g. Russell, Funke, Knott, & Strang, 2012). However, more general coding schemes can also be used, such as Entin and Entin (2001) and can help facilitate analyses such as anticipation ratios (overall, action, and information). These ratios can provide useful metrics for understanding team cognition by indicating whether teams spend more time anticipating the needs of their team or have to request their needs specifically.

Research Implications

Many of these measures have been validated in both laboratory and field contexts, however they have limited use and almost no validation within the context of cyber. In methods such as SAGAT, there must be a critical understanding of the environment such that an expert can extract key information requirements that map to Level 1, 2, and 3 SA (Endsley, 1988). The question of what is important for cyber SA is still open and important (Bardford et al., 2010). Several researchers have attempted to provide frameworks (e.g. D'Amica et al., 2005; D'Amico & Whitley, 2008; Champion et al., 2012; Gutzwiller et al., 2015, 2016), however many of these serve as one-off characterizations of a particular mission set. From a naturalistic viewpoint, a subject matter expert can articulate what they think is important in a task, but without a common understanding and framework for mapping, the results are not generalizable or comparable across studies, making the measure only useful within an individual assessment. Continued research should focus on developing frameworks that support previous work in team measurement so that findings across studies can be compared, allowing for a broader understanding of the role of team knowledge, attitudes, behaviors, and performance in cybersecurity tasks.

Another critical detail that must be discussed is the issue of ecological validity and applicability. While many of these measures have been tied to team performance in the lab and in several contexts, the unique nature of the cyber environment may have implications for how an instrument is used or its outcome. For example, an instrument like SAGAT is dependent on online knowledge of specific task critical information. In cyber, it is often the case that analysts rely on information repositories or other distributed cognitive artifacts. In this situation, it is not exactly critical to know and possess specific pieces of information, but rather, to know and understand where it is located, resonating with previous work on transactive memory (Wegner et al., 1991) and distributed cognition (Hutchins, 2006). This type of knowledge, of where to find information, can have similar performance effects as actually possessing the information and can also be used as an indicator of complete situational awareness (Sparrow, Liu, & Wegner, 2011; Stanton et al., 2006). Future research should conduct studies to better understand how previously validated measurements and instruments translate within cyber environments. This will help us better understand the greater theoretical underpinnings of each measurement and help produce generalizable instruments that can be used in cyber team research.

Table 6.4 provides a summarization of the exemplar challenges and associated research directions for team measurement in cybersecurity, as well as relevant citations from both cyber and team cognition literature.

Practical Implications

Many of the measures and metrics discussed in this section were designed and are often used in laboratory or controlled exercise settings. While they may not be ready for field use, cyber practitioners should begin to leverage previous work discussed here and elsewhere in collecting human-centered data during their assessments. Tool developers should be using situation awareness instruments to ensure that their users are able to obtain or locate the critical information in their tool, and use their findings to make interface updates and changes. Similarly, exercise developers, whether

TABLE 6.4
Summary of Challenges and Research Directions for Team Measures in Cybersecurity

Challenge	Research Directions	Relevant Citations
Performance in cyber is considered mostly from the technological perspective	• Triangulation across multiple measures (including technological) to develop more holistic and robust team performance measures	Baker and Salas (1997) Cannon-Bowers and Salas (2001) DeChurch and Mesmer-Magnus (2010) Wellens and Ergener (1988)
Current measurement techniques are often provided after the fact and provide little insight into team processes, procedures, or other intangible elements of teamwork.	• Correlation and development of team affect, stress, and workload scales to tangible outcomes in cybersecurity • Development (or extension of existing) of cyber team viability and cohesion instruments	Beal et al. (2003) Helton et al. (2014) Helton and Garland (2006) Lyons et al. (2011) Mathieu et al. (2008) Riggs and Knight (1994) Schwarz (2000)
Lack of ecological validity and applicability to cyber environments for existing measures	• Validation of existing instruments in high-fidelity environments (i.e. cyber exercises) • Development and validation of subject matter expert derived instruments (i.e. SAGAT) for general cyber tasks	Matthews and Beal (2002) Mazzocco et al. (2009) Patriciu and Furtuna (2009) Sommestad and Hallberg (2012) Sparrow et al. (2011)

for competition or training, should use the deluge of knowledge across the team literature and human sciences to begin to implement more rich metrics into their events. Cyber events could be a great opportunity to team with human factors and psychology researchers to help better understand cyber analyst processes and behaviors, and in turn, they could help the exercise developers with the design of these activities. As the research in these measurements becomes more mature and less invasive, a discussion on how to best implement into operations for the purpose of monitoring and organizational efficacy can then begin.

FUTURE DIRECTIONS FOR TEAM COGNITION IN CYBER

In this chapter, we have aimed to provide an overview of the current challenges for teamwork and team cognition that exist within cyber operations and security. However, as cyber is a digital domain, it also exists on the forefront of technology adoption. With this continued evolution, the issues of AI-human interaction and human-machine teaming are a growing challenge, with which many of the issues discussed in this chapter will overlap. As technology and the cyber environment continue to evolve, the reliance and presence of advanced and intelligent technology will also continue to grow. While we will most likely never reach a point in time in which technology will fully replace the human, technology working alongside humans is becoming more of a reality. With a move towards human-machine teaming, new research questions emerge, which team cognition may be especially suited to address. Many of the behaviors discussed in this chapter on how humans interact with others, such as team compositions, hand-offs, information sharing, etc., will still exist, however, between a human and a machine. This raises new questions about the role of team cognition in these environments: Is a human working with a machine a team? Does their cognitive structure exist as an individual or a team? What are the dynamics that exist between the collaboration? Before we can address these, we must first take a step back and understand what human-machine teaming is and identify paths for future exploration.

While human-machine teaming may be the new "hot" term, coupled with artificial intelligence, researchers have been discussing similar topics for well over a decade under the names automation and autonomy. Research recognizes automation across a spectrum, which ranges from the human doing everything to the computer being completely autonomous (Miller & Parasuraman, 2003). Much previous work does not necessarily deal with the autonomations or autonomous systems possessing "intelligence," however many of the same issues that plague human-machine teaming, such as trust, are equally as pervasive.

While we are not at the point where we have a full array of intelligent agents working alongside humans in cyber, within this spectrum, cyber has long relied on autonomous systems. Among the many types of automation in cyber, there are systems that automatically prioritize alerts based on a set of rules, systems that analyze files for potential malware, and algorithms that detect anomalous activity based on packets, connections, log-on events, and other data streams. There is also ongoing work to automate the processes of analysts through playbook-type workflow tools (Applebaum, Johnson, Limiero, & Smith, 2018), however, in most cases systems are providing information to the user and the user needs to review the results and decide on and implement an alternative action. The systems are providing recommendations but human and machine are not currently working as a team.

As these advance and mature, human-machine teaming can lead to improved performance if machines provide assistance in areas that are cognitively demanding for humans. One area is by improving the performance of novices by guiding them within a tool. Silva, Emmanuel, McClain, Matzen, and Forsythe (2015) had both novice and experts complete a forensic puzzle game, in which they had to answer questions regarding snapshots from different cyber tools. Results of the study were preliminary, however, using eye tracking they found that novices took longer to find the information of interest compared to experts. Further research is needed on features that can be added to tools to improve the performance of analysts with less experience. Automation and human-machine teaming could also help in areas that are difficult for all experience levels such as in identifying infrequent events. Sawyer and Hancock (2018) conducted an experiment in which participants interacted with an email testbed, in which they had to detect malicious emails. The rate at which malicious events occurred was of one of three levels: 1%, 5%, or 20% of the time. They found that response accuracy and response time was lowest for malicious events that occurred only 1% of the time compared to 5% and 20%. This is an exemplar where an AI teammate can be useful in augmenting human performance.

The cyber environment has unlimited potential for such capabilities and human-machine teaming. Cyber analysts need to integrate information from multiple data streams over wide time windows. Turner and Miller (2017) stated that studies are needed in which performance is examined when automation is used for intuitive processes, which are processes that do not involve working memory, and analytic processes, which are serial and affect working memory, in order to understand how different levels of automation affect an individual's ability to fuse information received. Without the appropriate assistance, it may be difficult for an analyst to piece together information from different analytic tools to identify malicious events. Having an agent that could pivot between tools and correlate information could aid the work of a human analyst. Based on the information gathered, there is also the potential for automating courses of action to further reduce the workload of human analysts. Russel and Norvig (2013) defined a taxonomy of agents that determine a course action, including simple reflex agents, which use only current information to make decisions, model-based agents, which infer actions based on model outcomes/predictions, goal-based agents, which determine an action in order to contribute to an overall goal, utility-based agents, which determine a course of action based on its desirability, and learning-based agents, which determine a course of action based on passed outcomes.

For a human-machine team to be most effective, the machine or agent needs to be trusted and found acceptable by its human counterpart. While a machine is able to respond and provide information instantaneously, this might not be beneficial as the timing of automated activities needs to be appropriate. If a machine presents information too quickly, a human analyst will be unable to interpret the results and decide on a response fast enough or will possibly reject the results as incomplete. If the machine though takes too long to complete its computations, the human analyst might ignore the machine agent, interpreting and deciding on courses of action on their own (Goodman, Miller, Rusnock, & Bindewald, 2016). Another important factor for the acceptance and use of automated capabilities is trust. Schaefer, Chen, Szalma, and Hancock (2016) conducted a meta-analysis of 30 studies to examine trust in automation and moderating factors. Emotive factors such as attitudes, comfort, and satisfaction with automation were the human traits that had the strongest effect size on the development of trust. Automation capability and behavior were the most important machine traits that affected trust. Hillsheim et al. (2017) found that trust in automation is also dependent on workload, with a higher workload associated with lack of trust of an agent. An agent may be viewed as untrustworthy if an individual's workload increases as a result of the agent. Lyons and Stokes (2012) found that trust and reliance in information is also modified by risk. They examined reliance on machine versus human when conflicting information is received for different levels of risk. Within the task they had to decide which route to send a military convoy based on route parameters, a tool that provided information on historical threats, or a human aid that provided a report based on intelligence. They found that high risk resulted in less reliance on the human aid. Studies examining the use of automation when making judgments during high- and low-risk cyber operational environments needs to be evaluated.

Like the discussion on performance for regular teams, performance of human-machine teams will need to be evaluated to determine if and how different types of automation improves cyber incident response. Damacharla, Javaid, Gallimore, and Devabhaktuni (2018) defined three metrics for benchmarking performance for human-machine teams. One metric is productive time, which is the total time of both autonomous and manual operation. The second metric was cohesion, which is a measure of how well a group remains united when completing tasks. One method for measuring cohesion is communication patterns. The final metric is the number of interventions needed to correct errors made by the machine. These metrics have not been studied in-depth for human-machine teams and testing and validation of all three metrics as well as development of additional performance metrics is needed.

This section aimed to provide an initial overview of automation and human-machine teaming literature, not an exhaustive list. We have provided initial potential directions for applying team cognition and automation to cybersecurity, but there is unlimited potential for researchers to address. While the reality of a fully intelligent team member may still be years away, it is on the horizon, and there are critical questions that must be addressed. Human factors and team cognition researchers should work to develop experimental protocols and methodologies to begin to understand how including an intelligent agent or machine into a team impacts our fundamental understanding of team cognition.

CONCLUSION

Cybersecurity, while growing in momentum, still remains a relatively unexplored frontier for human factors and team cognition scientists. In this paper, our goal was to begin to provide context-linking team research with current issues in cybersecurity to help steer future research, and provide information to practitioners who may want to leverage extant literature in their jobs. Based on an understanding of the cyber domain, we identified cyber SA, role definition and staffing hand-offs, and team measurement as exemplar topics which team cognition was especially suited to address. For each topic, we defined and outlined the problem space in relation to cybersecurity, discussed how team cognition may apply to it, and then discussed the implications for research and practice. Throughout this review we highlighted that improved processes and capabilities are needed in order to share information more effectively and improve SA within and across geographically distributed teams.

We also discussed a new frontier in team science which has numerous implications to cybersecurity: human-machine teaming. Processes within cybersecurity such as anomaly or malware detection are being automated, however there is little research to date on how a human and synthetic agent should work together within the cyber domain to reduce the workload of analysts and improve the detection of malicious activity. Moving forward, researchers and practitioners should continue to work together to help apply previous work in team cognition to the cyber environment, as well as continue to push the state of understanding in a new context to help improve our understanding of human interaction and begin to understand the implications of including intelligent machines into teams.

REFERENCES

Abramis, D. J. (1994). Work role ambiguity, job satisfaction, and job performance: Meta-analyses and review. *Psychological Reports*, *75*, 1411–1433.

Applebaum, A., Johnson, S., Limiero, M., & Smith, M. (2018, June). Playbook oriented cyber response. In *2018 National Cyber Summit (NCS)* (pp. 8–15). Huntsville, AL, USA: IEEE.

Asgarli, E., & Burger, E. (2016). Semantic ontologies for cyber threat sharing standards. In *IEEE symposium on technologies for homeland security*. Waltham, MA USA: IEEE.

Austin, J. R. (2003). Transactive memory in organizational groups: The effects of content, consensus, specialization, and accuracy on group performance. *Journal of Applied Psychology*, *88*(5), 866–878.

Baker, D. P., & Salas, E. (1997). Principles for measuring teamwork: A summary and look toward the future. In *Team performance assessment and measurement: Theory, methods, and applications* (pp. 331–355). Mahwah, N.J: Lawrence Erlbaum Associates.

Barford, P., Dacier, M., Dietterich, T. G., Fredrikson, M., Giffin, J., Jajodia, S., . . . Ou, X. (2010). Cyber SA: Situational awareness for cyber defense. In *Cyber situational awareness* (pp. 3–13). Boston, MA: Springer.

Bass, T. (2000). Intrusion detection systems and multisensor data fusion. *Communications of the ACM*, *43*(4), 99–105.

Batenburg, R., van Walbeek, W., & in der Mauer, W. (2013). Belbin role diversity and team performance is there a relationship? *Journal of Management Development*, *32*(8), 901–913.

Beal, D. J., Cohen, R. R., Burke, M. J., & McLendon, C. L. (2003). Cohesion and performance in groups: A meta-analytic clarification of construct relations. *Journal of Applied Psychology*, *88*(6), 989–1004.

Beauchamp, M. R., Bray, S. R., Eys, M. A., & Carron, A. V. (2002). Role ambiguity, role efficacy, and role performance: Multidimensional and mediational relationships within interdependent sport teams. *Group Dynamics: Theory, Research, and Practice*, *6*(3), 229–242.

Belbin, M. (1993). *Team roles at work.* Oxford: Butterworth-Heinemann.

Ben-Asher, N., & Gonzalez, C. (2015). Effects of cybersecurity knowledge on attack detection. *Computers in Human Behavior, 48*, 51–61.

Boyce, M. W., Duma, K. M., Hettinger, L. J., Malone, T. B., Wilson, D. P., & Lockett-Reynolds, J. (2011). *Human performance in cybersecurity a research agenda.* Paper presented at the Proceedings of the Human Factors and Ergonomics Society Annual Meeting, Las Vegas, NV.

Brannick, M. T., Roach, R. M., & Salas, E. (1993). Understanding team performance: A multimethod study. *Human Performance, 6*(4), 287–308.

Buchler, N., Rajivan, P., Marusich, L. R., Lightner, L., & Gonzalez, C. (2018). Sociometrics and observational assessment of teaming and leadership in a cybersecurity defense competition. *Computers & Security, 73*, 114–136.

Caltagirone, S., Pendergast, A., & Betz, C. (2013). *The diamond model of intrusion analysis (ADA586960).* Retrieved from Defense Technology Innovation Center www.dtic.mil/docs/citations/ADA586960.

Campbell, S. G., O'Rouke, P., & Bunting, M. F. (2015). *Identifying dimensions of cyber aptitude: The design of cyber aptitude and talent assessment.* Paper presented at the Proceedings of the Human Factors and Ergonomics Society Annual Meeting.

Cannon-Bowers, J. A., & Salas, E. (1998). Team performance and training in complex environments: Recent findings from applied research. *Current Directions in Psychological Science, 7*(3), 83–87.

Cannon-Bowers, J. A., & Salas, E. (2001). Reflections on shared cognition. *Journal of Organizational Behavior, 22*, 195–202.

Carron, A. V., Widmeyer, W. N., & Brawley, L. R. (1985). The development of an instrument to assess cohesion in sport teams: The group environment questionnaire. *Journal of Sport Psychology, 7*(3), 244–266.

Champion, M., Rajivan, P., Cooke, N. J., & Jariwala, S. (2012). *Team-based cyber defense analysis.* Paper presented at the IEEE International Multi-Disciplinary Conference on Cognitive Methods in Situation Awareness and Decision Support (CogSIMA), New Orleans, LA.

Chappelle, W., McDonald, K., Christensen, J., Prince, L., Goodman, T., Thompson, W., & Hayes, W. (2013). *Sources of occupational stress and prevalence of burnout and clinical distress among U.S. Air Force cyber warfare operators.* ARFL Report: ARFL-SA-WP-TR-2013-006.

Cisco. (2017). *Annual cybersecurity report.* Retrieved from www.cisco.com/go/acr2017.

Cohen, S. G., & Bailey, D. E. (1997). What makes teams work: Group effectiveness research from the shop floor to the executive suite. *Journal of Management, 23*(3), 239–290.

Cooke, N. J., Champion, M., Rajivan, P., & Jariwala, S. (2013). Cyber situation awareness and teamwork. *ICST Trans. Security Safety, 1*(2), e5.

Cooke, N. J., Gorman, J. C., Myers, C. W., & Duran, J. L. (2013). Interactive team cognition. *Cognitive Science, 37*(2), 255–285.

Cooke, N. J., Salas, E., Cannon-Bowers, J. A., & Stout, R. J. (2000). Measuring team knowledge. *Human Factors, 42*(1), 151–173.

Cooper, G. E., & Harper Jr, R. P. (1969). *The use of pilot rating in the evaluation of aircraft handling qualities* (No. AGARD-567). Neuilly-Sur-Seine, France: Advisory Group for Aerospace Research and Development.

Cota, A. A., Evans, C. R., Dion, K. L., Kilik, L., & Longman, R. S. (1995). The structure of group cohesion. *Personality and Social Psychology Bulletin, 21*(6), 572–580.

Damacharla, P., Javaid, A., Gallimore, J., & Devabhaktuni, V. (2018). Common metrics to benchmark human-machine team (HMT): A review. *IEEE Access, 6*, 38637–38655.

D'Amico, A., Tesone, D., Whitley, K., O'Brien, B., Smith, M., & Roth, E. (2005). *Understanding the cyber defender: A cognitive task analysis of information assurance analysts*. Report No. CSA-CTA-ll. Secure Decisions. Funded by ARDA and DOD.

D'Amico, A., & Whitley, K. (2008). The real work of computer network defense analysts. In *VizSEC 2007* (pp. 19–37). Berlin and Heidelberg: Springer.

DeChurch, L. A., & Mesmer-Magnus, J. R. (2010). Measuring shared team mental models: A meta-analysis. *Group Dynamics: Theory, Research, and Practice*, *14*(1), 1–14.

DeStefano, C., Lachevet, K., & Carozzoni, J. (2008). *Distributed planning in a mixed-initiative environment*. Rome, NY: Air Force Research Laboratory, Systems and Information Interoperability Branch.

Dillenbourg, P., Baker, M. J., Blaye, A., & O'Malley, C. (1996). The evolution of research on collaborative learning. In *Learning in humans and machine: Towards an interdisciplinary learning science* (pp. 189–211). Emerald Group Publishing Limited Bingley, UK.

Dinsmore, D. L., Alexander, P. A., & Loughlin, S. M. (2008). Focusing the conceptual lens on metacognition, self-regulation, and self-regulated learning. *Educational Psychology Review*, *20*(4), 391–409.

Driskell, J. E., & Salas, E. (1991). Group decision making under stress. *Journal of Applied Psychology*, *76*(3), 473–478.

Driskell, J. E., Salas, E., & Johnston, J. (1999). Does Stress Lead to a loss of team perspective? *Group Dynamics: Theory, Research, and Practice*, *3*(4), 291–302.

Dyer, J. L. (1984). Team research and team training: A state-of-the-art review. *Human Factors Review*, *26*, 285–323.

Edmondson, A. (1999). Psychological safety and learning behavior in work teams. *Administrative Science Quarterly*, *44*(2), 350–383.

Edwards, B. D., Anthony Day, E., Arthur Jr, W., & Bell, S. T. (2006). Relationships among team ability composition, team mental models and team performance. *Journal of Applied Psychology*, *91*(3), 727–736.

Einhorn, H. J., & Hogarth, R. M. (1978). Confidence in judgement: Persistence of the illusion of validity. *Psychological Review*, *85*(5), 395–416.

Endsley, M. R. (1988). Situation awareness global assessment technique (SAGAT). In *Aerospace and electronics conference, 1988. NAECON 1988. Proceedings of the IEEE 1988 National Conference* (pp. 789–795). Dayton, OH: IEEE.

Endsley, M. R. (1989). *Situation awareness in an advanced strategic mission (No. NOR DOC 89–32)*. Hawthorne, CA: Northrop Corporation.

Endsley, M. R. (1995). Toward a theory of situation awareness in dynamic systems. *Human Factors: The Journal of the Human Factors and Ergonomics Society*, *37*(1), 32–64.

Endsley, M. R., & Connors, E. S. (2014). Foundation and challenges. In *Cyber defense and situational awareness* (pp. 7–27). Cham: Springer.

Endsley, M. R., & Rodgers, M. D. (1994). Situation awareness information requirements analysis for en route air traffic control. In *Proceedings of the Human Factors and Ergonomics Society annual meeting* (Vol. 38, No. 1, pp. 71–75). Los Angeles, CA: SAGE Publications.

Endsley, M. R., Selcon, S. J., Hardiman, T. D., & Croft, D. G. (1998). A comparative analysis of SAGAT and SART for evaluations of situation awareness. In *Proceedings of the Human Factors and Ergonomics Society annual meeting* (Vol. 42, No. 1, pp. 82–86). Los Angeles, CA: SAGE Publications.

Entin, E. E., & Entin, E. B. (2001). Measures for evaluation of team processes and performance in experiments and exercises. In *Proceedings of the 6th international command and control research and technology symposium* (pp. 1–14). Annapolis, MD, USA: International C2 Institute.

Espinosa, J. A., Nan, N., & Carmel, E. (2015). Temporal distance, communication patterns, and task performance in teams. *Journal of Management Information Systems, 32*(1), 151–191.

Fairfax, T., Laing, C., & Vickers, P. (2015). Network situational awareness: Sonification and visualization in the cyber battlespace. In *Handbook of research on digital crime, cyberspace security, and information assurance* (pp. 334–349). Hershey, PA: IGI Global.

Fanelli, R., & Conti, G. (2012, June). A methodology for cyber operations targeting and control of collateral damage in the context of lawful armed conflict. In *2012 4th international Conference on Cyber Conflict (CYCON 2012)* (pp. 1–13). Tallinn, Estonia: IEEE.

Faraj, S., & Xiao, Y. (2006). Coordination in fast-response organizations. *Management Science, 52*(8), 1155–1169.

Federal Bureau of Investigation (FBI). (2016). *Internet crimes report 2016*. Retrieved from https://pdf.ic3.gov/2016_IC3Report.pdf

Fink, G., North, C. L., Endert, A., & Rose, S. (2009). *Visualizing cybersecurity: Usable workspaces*. Paper presented at the 6th International Workshop on Visualization for Cyber Security. Atlantic City, NJ.

Gersh, J. R., & Bos, N. (2014). *Cognitive and organizational challenges of big data in cyber defense*. Paper presented at the 1st workshop on Human-Centered Big Data Research. Raleigh, NC.

Goodall, J. R., D'Amico, A., & Kopylec, J. K. (2009, October). Camus: Automatically mapping cyber assets to missions and users. In *MILCOM 2009–2009 IEEE military communications conference* (pp. 1–7). Atlantic City, NJ: IEEE.

Goodman, T., Miller, M. E., Rusnock, C. F., & Bindewald, J. (2016). Timing within human agent interaction and its effects on team performance and human behavior. In *Proceedings of the IEEE international multi-disciplinary conference on Cognitive Methods in Situation Awareness and Decision Support, CogSIMA 2016* (pp. 35–41). San Diego, CA: IEEE.

Gutzwiller, R. S., Fugate, S., Sawyer, B. D., & Hancock, P. A. (2015). The human factors of cyber network defense. In *Proceedings of the Human Factors and Ergonomics Society annual meeting* (Vol. 59, No. 1, pp. 322–326). Los Angeles, CA: SAGE Publications.

Gutzwiller, R. S., Hunt, S. M., & Lange, D. S. (2016). A task analysis toward characterizing cyber-cognitive situation awareness (CCSA) in cyber defense analysts. In *Proceedings of the IEEE international multi-disciplinary conference on Cognitive Methods in Situation Awareness and Decision Support, CogSIMA 2016* (pp. 14–20). San Diego, CA: IEEE.

Hamilton, K., Mancuso, V., Mohammed, S., Tesler, R., & McNeese, M. (2017). Skilled and unaware: The interactive effects of team cognition, team metacognition, and task confidence on team performance. *Journal of Cognitive Engineering and Decision Making, 11*(4), 382–395.

Hansen, M. T. (1999). The Search-Transfer Problem: The Role of Weak Ties in Sharing Knowledge across Organization Subunits. *Administrative Science Quarterly, 44*(1), 82–111. https://doi.org/10.2307/2667032

Harrald, J., & Jefferson, T. (2007). Shared situational awareness in emergency management mitigation and response. In *System sciences, 2007. HICSS 2007* (pp. 23–23). Waikoloa, HI: IEEE.

Hart, S. G., & Staveland, L. E. (1988). Development of NASA-TLX (Task Load Index): Results of empirical and theoretical research. *Advances in Psychology, 52*, 139–183.

Hayes-Roth, B., Hayes-Roth, F., Rosenchein, S. J., & Cammarata, S. (1979). *Modeling planning as an incremental, opportunistic process* (No. RAND/N-1178-ONR). Santa Monica, CA: RAND Corp.

Hellar, D. B., & McNeese, M. (2010, September). NeoCITIES: A simulated command and control task environment for experimental research. In *Proceedings of the Human*

Factors and Ergonomics Society annual meeting (Vol. 54, No. 13, pp. 1027–1031). Los Angeles, CA: SAGE Publications.

Helton, W. S., Funke, G. J., & Knott, B. A. (2014). Measuring workload in collaborative contexts: Trait versus state perspectives. *Human Factors*, *56*(2), 322–332.

Helton, W. S., & Garland, G. (2006). Short stress state questionnaire: Relationships with reading comprehension and land navigation. In *Proceedings of the Human Factors and Ergonomics Society annual meeting* (Vol. 50, No. 17, pp. 1731–1735). Los Angeles, CA: SAGE Publications.

Henschel, D., Deckard, G. M., Lufkin, B., Buchler, N., Hoffman, B., Rajivan, P., & Collman, S. (2016). Predicting proficiency in cyber exercises. In *Proceedings of MILCOM 2016–2016 IEEE military communications conference*. Baltimore, MD: IEEE.

Herjavec Group. (2017). *Cybersecurity jobs report*. A Special Report from the Editors at Cybersecurity Ventures. Retrieved from https://cybersecurityventures.com/jobs/.

Hui, P., Bruce, J., Fink, G., Gregory, M., Best, D., McGrath, L., & Endert, A. (2010). *Towards efficient collaboration in cybersecurity*. Paper presented at the 2010 International Symposium on Collaborative Technologies and Systems (CTS), Chicago, IL.

Hutchins, E. (2006). The distributed cognition perspective on human interaction. *Roots of human sociality: Culture, cognition and interaction*, 1, 375.

ISC^2. (2017). *Global information security workforce study: Benchmarking workforce capacity and response to cyber risk*. Retrieved from https://iamcybersafe.org/gisws/.

Jiang, S. Y., Murphy, A., Heitkemper, E. M., Hum, S., Kaufman, D. R., & Mamykina, L. (2017). Impact of an electronic handoff documentation tool on team shared mental models in pediatric critical care. *Journal of Biomedical Informatics*, *69*, 24–32.

Johnston, J. H., Smith-Jentsch, K. A., & Cannon-Bowers, J. A. (1997). Performance measurement tools for enhancing team decision making. In M. T. Brannick, E. Salas, & C. Prince (Eds.), *Team performance assessment and measurement: Theory, research, and applications* (pp. 311–330). Hillsdale, NJ: Erlbaum.

Joint Commission. (2012). *Joint commission center for transforming healthcare releases targeted solutions tool for hand-off communications*. Retrieved from www.jointcommission.org/assets/1/6/TST_HOC_Persp_08_12.pdf

Joint Publications. (2018). *Joint Publications 3–12 cyberspace operations*. Retrieved from www.jcs.mil/Doctrine/Joint-Doctrine-Pubs/3-0-Operations-Series/.

Jones, D. G., & Endsley, M. R. (1996). Sources of situation awareness errors in aviation. *Aviation, Space, and Environmental Medicine*, *67*(6), 507–512.

Kaber, D. B., Perry, C. M., Segall, N., McClernon, C. K., & Prinzel III, L. J. (2006). Situation awareness implications of adaptive automation for information processing in an air traffic control-related task. *International Journal of Industrial Ergonomics*, *36*(5), 447–462.

Kahn, R. L., Wolfe, D. M., Quinn, R. P., Snoek, J. D., & Rosenthal, R. A. (1964). *Conflict and ambiguity: Studies in organizational roles and individual stress*. Oxford, UK: John Wiley.

Keebler, J. R., Lazzara, E. H., Patzer, B. S., Palmer, E. M., Plummer, J. P., Smith, D. C., . . . Riss, R. (2016). Meta-analyses of the effects of standardized handoff protocols on patient, provider and organizational outcomes. *Human Factors*, *58*(8), 1187–1205.

Kendall, D. L., & Salas, E. (2004). Measuring team performance: Review of current methods and consideration of future needs. *The Science and Simulation of Human Performance*, *5*, 307–326.

Klimoski, R., & Mohammed, S. (1994). Team mental model: Construct or metaphor? *Journal of Management*, *20*(2), 403–437.

Knott, B. A., Mancuso, V. F., Bennett, K., Finomore, V., McNeese, M., McKneely, J. A., & Beecher, M. (2013). Human factors in cyber warfare: Alternative perspectives.

In *Proceedings of the Human Factors and Ergonomics Society annual meeting* (Vol. 57, No. 1, pp. 399–403). Los Angeles, CA: SAGE Publications.

Lyons, J. B., Funke, G. J., Nelson, A., & Knott, B. A. (2011). Exploring the impact of cross-training on team process. In N. A. Stanton (Ed.), *Trust in military teams* (pp. 49–70). Aldershot, UK: Ashgate.

Lyons, J. B., & Stokes, C. K. (2012). Human-human reliance in the context of automation. *Human Factors*, *54*(1), 112–121.

MacDonald Jr, A. P., Kessel, V. S., & Fuller, J. B. (1972). Self-disclosure and two kinds of trust. *Psychological Reports*, *30*(1), 143–148.

Mañas, M. A., Díaz-Fúnez, P., Pecino, V., López-Liria, R., Padilla, D., & Aguilar-Parra, J. M. (2018). Consequences of team job demands: Role ambiguity climate, affective engagement, and extra-role performance. *Frontiers in Psychology*, *8*, 2292.

Mancuso, V. F., Christensen, J. C., Cowley, J., Finomore, V., Gonzalez, C., & Knott, B. (2014). Human factors in cyber warfare II: Emerging perspectives. In *Proceedings of the Human Factors and Ergonomics Society annual meeting* (Vol. 58, No. 1, pp. 415–418). Los Angeles, CA: SAGE Publications.

Mancuso, V. F., Minotra, D., Giacobe, N., McNeese, M., & Tyworth, M. (2012). idsNETS: An experimental platform to study situation awareness for intrusion detection analysts. In *2012 IEEE international multi-disciplinary conference on Cognitive Methods in Situation Awareness and Decision Support (CogSIMA)* (pp. 73–79). Chicago, IL: IEEE.

Marlow, S. L., Lacerenza, C. N., Paoletti, J., Burke, C. S., & Salas, E. (2018). Does team communication represent a one-size-fits-all approach?: A meta-analysis of team communication and performance. *Organizational Behavior and Human Decision Processes*, *144*, 145–170.

Mathieu, J. E., Maynard, M. T., Rapp, T., & Gilson, L. (2008). Team effectiveness 1997–2007: A review of recent advancements and a glimpse into the future. *Journal of Management*, *34*(3), 410–476.

Mathieu, J. E., Tannebaum, S. I., Donsbach, J. S., & Alliger, G. M. (2014). A review and integration of team composition models: Moving toward a dynamic and temporal Framework. *Journal of Management*, *40*(1), 130–160.

Matthews, G., Campbell, S. E., Desmond, P. A., Huggins, J., Falconer, S., & Joyner, A. (1999). Assessment of task-induced state change: Stress, fatigue and workload components. *Automation technology and human performance: Current research and trends*, 199–203.

Matthews, M. D., & Beal, S. A. (2002). *Assessing situation awareness in field training exercises*. Military Academy, West Point, NY: Office of Military Psychology and Leadership.

Mayer, R. C., & Davis, J. H. (1999). The effect of the performance appraisal system on trust for management: A field quasi-experiment. *Journal of applied psychology*, *84*(1), 123.

Mazzocco, K., Petitti, D. B., Fong, K. T., Bonacum, D., Brookey, J., Graham, S., . . . Thomas, E. J. (2009). Surgical team behaviors and patient outcomes. *The American Journal of Surgery*, *197*(5), 678–685.

McCallam, D. H., Frazier, P. D., & Savold, R. (2017, July). Ubiquitous connectivity and threats: Architecting the next generation cybersecurity operations. In *2017 IEEE 7th annual international conference on CYBER technology in automation, control, and intelligent systems (CYBER)* (pp. 1506–1509). Kaiulani, HI: IEEE.

McGuinness, B., & Foy, L. (2000). A subjective measure of SA: The Crew Awareness Rating Scale. In D. B. Kaber and M. R. Endsley (Eds.), *Human performance, situation awareness and automation: User centered design for the new millennium*. Atlanta, GA: SA Technologies.

McNeese, M. D., Connors, E. S., Jones, R. E., Terrell, I. S., Jefferson Jr, T., Brewer, I., & Bains, P. (2005). Encountering computer-supported cooperative work via the living lab:

Application to emergency crisis management. In *Proceedings of the 11th international conference of human-computer interaction*.
McNeese, M. D., Mancuso, V. F., McNeese, N. J., Endsley, T., & Forster, P. (2013). *Using the living laboratory framework as a basis for understanding next-generation analyst work*. Paper presented at the SPIE Defense, Security, and Sensing, Baltimore, MD.
McNeese, M. D., Mancuso, V. F., McNeese, N. J., Endsley, T., & Forster, P. (2014). An integrative simulation to study team cognition in emergency crisis management. In *Proceedings of the Human Factors and Ergonomics Society annual meeting* (Vol. 58, No. 1, pp. 285–289). Los Angeles, CA: Sage Publications.
Miller, C. A., & Parasuraman, R. (2003). Beyond levels of automation: An architecture for more flexible human-automation collaboration. In *Proceedings of the Human Factors and Ergonomics Society 47th annual meeting*. Denver, CO.
Mohammed, S., & Dumville, B. C. (2001). Team mental models in a team knowledge framework: Expanding theory and measurement across disciplinary boundaries. *Journal of Organizational Behavior, 22*(2), 89–106.
Newhouse, W., Keith, S., Scribner, B., & Witte, G. (2017). *National Initiative for Cybersecurity Education (NICE) cybersecurity workforce framework*. NIST Special Publication 800–181. Retrieved from https://doi.org/10.6028/NIST.SP.800-181.
O'Leary, M. B., & Cummings, J. N. (2007). The spatial, temporal, and configurational characteristics of geographic dispersion in teams. *MIS quarterly*, 433–452.
Oltsik, J. (2017). Cybersecurity skills shortage hurts security analytics, operations. *CSO*. Retrieved from https://search.proquest.com/docview/1922897602?accountid=12492.
Oshri, I., van Fenema, P., & Kotlarsky, J. (2008). Knowledge transfer in globally distributed teams: The role of transactive memory. *Information Systems Journal, 18*, 593–616.
Patriciu, V. V., & Furtuna, A. C. (2009). Guide for designing cybersecurity exercises. In *Proceedings of the 8th WSEAS international conference on E-activities and information security and privacy* (pp. 172–177). Stevens Point, WI: World Scientific and Engineering Academy and Society (WSEAS).
Paul, C. L., & Dykstra, J. (2017). Understanding operator fatigue, frustration, and cognitive workload in tactical cybersecurity operations. *Journal of Information Warfare, 16*(2), 1–11.
Paulsen, C., McDuffie, E., Newhouse, W., & Toth, P. (2012). NICE: Creating a cybersecurity workforce and aware public. *IEEE Security and Privacy, 10*(3), 76–79.
Rajivan, P., Champion, M., Cooke, N. J., Jariwala, S., Dube, G., & Buchanan, V. (2013, July). Effects of teamwork versus group work on signal detection in cyber defense teams. In *International conference on augmented cognition* (pp. 172–180). Berlin and Heidelberg: Springer.
Reid, G. B., & Nygren, T. E. (1988). The subjective workload assessment technique: A scaling procedure for measuring mental workload. In P. A. Hancock & N. Meshkati (Eds.), *Human mental workload* (pp. 185–218). Amsterdam: Elsevier.
Riggs, M. L., & Knight, P. A. (1994). The impact of perceived group success-failure on motivational beliefs and attitudes: A causal model. *Journal of Applied Psychology, 79*, 755–766.
Rodgers, M. (2017). *Human factors impacts in air traffic management*. Abingdon: Routledge.
Roscoe, A. H. (1987). *The practical assessment of pilot workload*. AGARD-AG-282. Neuilly-Sur-Seine, France: Advisory Group for Aerospace Research and Development.
Rousseau, R., Tremblay, S., Banbury, S., Breton, R., & Guitouni, A. (2010). The role of metacognition in the relationship between objective and subjective measures of situation awareness. *Theoretical Issues in Ergonomics Science, 11*(1–2), 119–130.
Russel, S., & Norvig, P. (2013). Intelligent agents. In *Artificial intelligence: A modern approach* (pp. 47–57). Upper Saddle River, NJ: Pearson Education Limited.

Russell, S. M., Funke, G. J., Knott, B. A., & Strang, A. J. (2012). *Recurrence quantification analysis used to assess team communication in simulated air battle management.* Paper presented at the Proceedings of the Human Factors and Ergonomics Society Annual Meeting.

Salas, E., Bowers, C. A., & Cannon-Bowers, J. A. (1995). Military team research: 10 years of progress. *Military Psychology, 7*(2), 55–75.

Salas, E., Cooke, N. J., & Rosen, M. A. (2008). On teams, teamwork, and team performance: Discoveries and developments. *Human Factors, 50*(3), 540–547.

Sawyer, B. D., & Hancock, P. A. (2018). Hacking the human: The prevalence paradox in cybersecurity. *Human Factors, 60*(5), 597–609.

Schaefer, K. E., Chen, J. Y., Szalma, J. L., & Hancock, P. A. (2016). A Meta-analysis of factors influencing the development of trust in automation: Implications for understanding autonomy in future systems. *Human Factors, 58*(3), 377–400.

Schwarz, N. (2000). Emotion, cognition, and decision making. *Cognition & Emotion, 14*(4), 433–440.

Shuffler, M. L., & Carter, D. R. (2018). Teamwork situated in multiteam systems: Key lessons learned and future opportunities. *American Psychologist, 73*(4), 390.

Silva, A., Emmanuel, G., McClain, J. T., Matzen, L., & Forsythe, C. (2015). *Measuring expert and novice performance within computer security incident response teams.* Lecture Notes in Computer Science Volume 9183, 2015, pp. 144–152, 9th International Conference on Augmented Cognition, AC 2015 held as part of 17th International Conference on Human Computer Interaction, HCI International 2015, Los Angeles, United States, 2 August 2015 through 7 August 2015.

Smith, D. R. (1999). *The effect of transactive memory and collective efficacy on aircrew performance.* Dayton, OH: Wright State University.

Sommestad, T., & Hallberg, J. (2012). Cyber security exercises and competitions as a platform for cybersecurity experiments. In *Nordic conference on secure IT systems* (pp. 47–60). Berlin and Heidelberg: Springer.

Sparrow, B., Liu, J., & Wegner, D. M. (2011). Google effects on memory: Cognitive consequences of having information at our fingertips. *Science, 333*(6043), 776–778.

Staheli, D., Mancuso, V., Harnasch, R., Fulcher, C., Chmielinski, M., Kearns, A., . . . Vuksani, E. (2016). Collaborative data analysis and discovery for cybersecurity. In *WSIW@ SOUPS.* Denver, CO: USENIX Association.

Stanton, N. A., Stewart, R., Harris, D., Houghton, R. J., Baber, C., McMaster, R., . . . Linsell, M. (2006). Distributed situation awareness in dynamic systems: Theoretical development and application of an ergonomics methodology. *Ergonomics, 49*(12–13), 1288–1311.

Star, S. L., & Griesemer, J. R. (1989). Institutional ecology, translations' and boundary objects: Amateurs and professionals in Berkeley's Museum of Vertebrate Zoology, 1907–39. *Social Studies of Science, 19*(3), 387–420.

Stewart, G. L. (2006). A meta-analytic review of relationships between team design features and team performance. *Journal of Management, 32*(1), 29–55.

Taylor, R. M. (2017). Situational Awareness Rating Technique (SART): The development of a tool for aircrew systems design. In *Situational awareness* (pp. 111–128). New York: Routledge.

Tsang, P. S., & Velazquez, V. L. (1996). Diagnosticity and multidimensional subjective workload ratings. *Ergonomics, 39*(3), 358–381.

Turner, K. L., & Miller, M. E. (2017). *The effect of automation and workspace design. Humans' ability to recognize patterns while fusing information.* Paper presented at IEEE Conferences on Cognitive and Computational Aspects of Situation Management (CogSIME), Savannah, GA.

Tyworth, M., Giacobe, N. A., Mancuso, V. F., & Dancy, C. D. (2012). *The distributed nature of cyber situation awareness*. Paper presented at the 2012 IEEE International Multi-Disciplinary Conference on Cognitive Methods in Situation Awareness and Decision Support, New Orleans, LA.

Tyworth, M., Giacobe, N. A., Mancuso, V. F., McNeese, M. D., & Hall, D. L. (2013). A human-in-the-loop approach to understanding situation awareness in cyber defense analysis. *EAI Endorsed Transactions on Security and Safety*, *1*(2), 1–10.

Uitdewilligen, S., & Waller, M. (2018). Information sharing and decision-making in multidisciplinary crisis management teams. *Journal of Organizational Behavior*, *39*(6), 731–748.

Van de Water, H., Ahaus, K., & Rozier, R. (2008). Team roles, team balance and performance. *Journal of Management Development*, *27*(5), 499–512.

Vidulich, M. A., & Hughes, E. R. (1991). Testing a subjective metric of situation awareness. In *Proceedings of the Human Factors and Ergonomics Society annual meeting* (Vol. 35, No. 18, pp. 1307–1311). Los Angeles, CA: SAGE Publications.

Volpe, C. E., Cannon-Bowers, J. A., Salas, E., & Spector, P. E. (1996). The impact of cross-training on team functioning: An empirical investigation. *Human Factors*, *38*(1), 87–100.

Waag, W. L., & Houck, M. R. (1994). Tools for assessing situational awareness in an operational fighter environment. *Aviation, Space, and Environmental Medicine*, *65*(5 Suppl), A13–A19.

Watson, D., & Clark, L. A. (1999). *The PANAS-X: Manual for the positive and negative affect schedule-expanded form*. Iowa City, IA: The University of Iowa.

Watson, D., Clark, L. A., & Tellegen, A. (1988). Development and validation of brief measures of positive and negative affect: The PANAS scales. *Journal of Personality and Social Psychology*, *54*(6), 1063–1070.

Wegner, D. M. (1995). A computer network model of human transactive memory. *Social Cognition*, *13*(3), 319–339.

Wegner, D. M., Erber, R., & Raymond, P. (1991). Transactive memory in close relationships. *Journal of Personality and Social Psychology*, *61*(6), 923–929.

Wellens, A. R. (1993). Group situation awareness and distributed decision making: From military to civilian applications. In N. J. J. Castellan (Ed.), *Individual and group decision making: Current issues* (pp. 267–287). Hillsdale, NJ: Lawrence Erlbaum Associates.

Wellens, A. R., & Ergener, D. (1988). The CITIES game: A computer-based situation assessment task for studying distributed decision making. *Simulation & Games*, *19*(3), 304–327.

Wickens, C. D. (2002). Situation awareness and workload in aviation. *Current Directions in Psychological Science*, *11*(4), 128–133.

Zetter, K. (2016, March 3). *Inside the cunning, unprecedented hack of Ukraine's power grid*. Retrieved September 20, 2019, from www.wired.com/2016/03/inside-cunning-unprecedented-hack-ukraines-power-grid/.

Zimmerman, C. (2014). *Ten strategies of a world-class cybersecurity operations center*. MITRE Corporate Communications and Public Affairs.

7 Distributed Cognition and Human-Co-Robot Manufacturing Teams

Issues in Design and Implementation

Lora A. Cavuoto and Ann M. Bisantz

CONTENTS

INTRODUCTION

Collaborative robots (co-robots) present a potentially transformative technology for small, medium, and large manufacturers because they enable manufacturing's desire for rapid, flexible automation capable of semi-manual/automated tasks. Manufacturing co-robots exploit the flexibility and creative aspects of (human) manual work with the efficiency and productivity of automation—if they are designed and integrated into manufacturing operations effectively.

Co-robots have been deployed across a range of manufacturing environments, with the largest implementations existing in the automotive, electronics, and medtech industries. In automotive manufacturing, for example, manufacturers have installed co-robots on assembly lines for high-precision tasks such as screw-driving, sealant dispensing, and plug insertion. Other applications include electronics inspection and carton packaging. A recent study by BMW and MIT showed that human-co-robot teams performing cooperative work reduced human idle time by about 80% and were more productive than human- or robot-only teams (Unhelkar, Lasota et al., 2018). Most current applications involve payloads in the range of 1 to 5 kg. Co-robot adaptability and affordability, with costs on the order of $50,000 to $100,000 depending on the type and payload requirements, mean that small and medium-sized enterprises (SMEs) have the ability to implement these systems. Scale-up of co-robot installations in manufacturing will continue, due to faster, lower cost deployment than traditional automation, the smaller required footprint of the technology, the potential reduction in ergonomic risk to operators, and an expected labor shortage of skilled manufacturing workers. However, effective scale-up requires a fundamental understanding of manufacturing-specific human-co-robot interaction. Failure to do so can be detrimental in terms of efficiency, productivity, flexibility, and quality, as has been the case with over half of the early introductions of advanced manufacturing technologies and cellular manufacturing (Chung, 1996; Charalambous, Fletcher, & Webb, 2015).

Defining Human-Co-Robot Teams

In designing for effective use, it is important to define the terminology related to co-robots and what distinguishes them from traditional industrial robots. A robot includes the robot arm and its means of control (both software and electronics control system) and a robot system includes the robot along with any accompanying end-effector and parts that are manipulated by the robot arm. Operators include all personnel that interact with the system, including production personnel as well as any maintenance and programming personnel.

In a traditional manufacturing environment with industrial automation, industrial robots are typically separated from the human operator by means of a cage or other physical barrier. When the robot is active, the operator is prevented from entering the blocked area, minimizing the risk of interference between the human and robot. For these systems, the human and robot coexist, operating at different times in separate workspaces. In a second type of robot operation, termed cooperation, a similar industrial robot is used in the same workspace as a human, however all human tasks are done when the robot is inactive, and robot functioning starts once the human is at a safe distance away. An example of cooperation includes tasks where the human operator is responsible for loading and unloading parts. In contrast, with collaborative operation, the human and co-robot share the same space and are active at the same time. As defined by the Robotic Industries Association (RIA; ANSI/RIA Technical Report R15.606–2016) and the International Organization of Standardization (ISO) Technical Specification (ISO/TS 15066, 2016; International Organization for Standardization, 2016; Robotics Industries Association, 2016), a collaborative robot is “a robot that can be used in a collaborative operation within a collaborative workspace” where collaborative operation is a “state in which a purposely designed robot system and an operator work within a collaborative workspace” and collaborative workspace is a “space within the operating space where the robot (including the workpiece) and a human can perform tasks concurrently during production operation.” A collaborative robot is defined by the task the robot is performing, the timing of that task with respect to the operator’s task, and the space in which the task is being performed, not by the robot itself. Note that collaborative robots, by these definitions, involve collaboration between people and robots occupying the same physical space (and typically, working together to manufacture the same physical artifact), contemporaneously. That is, the notion of collaborative robots, as defined and discussed in this chapter, does not refer to asynchronous collaboration or collaboration at a distance.

An important distinction between co-robots and traditional industrial robots is that co-robots are typically power and force limited, minimizing the risk of injury to the operator if contact occurs. Contact can be intended (through hand guidance), incidental (unintended impact between the operator and robot), or resulting from a failure of the co-robot. The co-robot has sensors and safety functions that detect the contact and stop. Allowable levels of contact force are based on the level of risk and likelihood of pain for the operator (Muttray, Melia et al., 2014; International Organization for Standardization, 2016). Christiernin (2017) provides a relevant classification framework for characterizing the level of interaction with co-robots, ranging from low levels with no collaboration (traditional robots that perform work in a separated, gated space from humans) and co-location only when the robot is idle or moving slowly under human command, to higher-level collaborative situations. These include, for example, humans and robots working in co-located spaces, but where the co-robots' sensors detect human movements and avoid contact, or human-guided robot movement. In most cases, co-robot movement is preplanned based on task requirements and knowledge of expected human movement paths, but adaptive learning is being integrated to support adaptability to unexpected situations. Adaptive learning is discussed later in this chapter. At the highest "collaboration" level, the humans and robots are "working together in the same physical space, solving problems together" and robots adapt their behaviors based on human activities, and may initiate actions.

The primary focus of much of the initial research and development work regarding the integration of co-robots into manufacturing environments has focused on the safety aspects of eliminating machine guarding with co-robot installation. Thus, important questions regarding the roles that co-robots make take on in human-co-robot teams and how both co-robots and teamwork will need to be designed to support successful human-co-robot teams still remain. In response, this chapter identifies theoretical frameworks drawn from human factors and distributed cognition which can inform the study and design of human-co-robot manufacturing teams, including situation awareness, communication, trust/reliability, and function allocation. The impacts of the teams on operator safety, production quality, and efficiency are also explored. Implications for system (both robot and work environment) design and operator training are presented.

THEORETICAL FRAMEWORKS FOR CONSIDERING HUMAN-CO-ROBOT MANUFACTURING TEAMS

Situation Awareness

A critical theoretical lens with which to view distributed human-co-robot teams engaged in the highest level of collaboration (Christiernin, 2017) is that of situation awareness. Endsley's (1995) three-level model of situation awareness can be applied: Level 1, awareness of dynamic system state variables; Level 2, understanding of the meaning of those variable values in context; and Level 3, prediction of the future state of those variables. For any manufacturing situation, these levels would apply to human awareness of relevant manufacturing process variables that make up the work task. Additionally, however, the introduction of a co-robot team member requires additional layers of awareness. Currently, the three most common applications of manufacturing co-robots are in machine tending (i.e. loading raw material or unloading finished parts from a machine like a press), pick and place (i.e. moving parts from one location to another, often with precision), and dispensing (e.g., depositing a bead of fluid adhesive). Consider a joint assembly task in which a co-robot retrieves a part from a 3-D printer and places a part in a fixture, the human operator visually inspects the part before manually installing a clip, and then advances the fixture to a second co-robot, which applies a bead of adhesive. From the human operator perspective, relative to the co-robot, levels of situation awareness in this case would translate to (1) knowing the position and directional velocity of the co-robot arms and actions being taken by the end-effectors for both co-robot arms, (2) understanding the task that is being performed by the robot (e.g., for the first co-robot, the tasks of moving to placing the part, releasing the part, or moving to retrieve the next part; for the second co-robot, tasks of preparing to dispense, dispensing along a path, or resetting to the start of the path), and (3) predicting the future actions and states of the arms (trajectory and end position) and effectors and the next tasks to be performed. Considering the co-robot to be a team member (rather than a system the human is controlling) requires a similar sense of awareness on the part of the co-robot itself. That is, to truly support collaboration as a distributed team, the co-robot should itself have awareness of the work task variables, particularly, in this example, whether or not the human operator has released the previous part (and is therefore ready to receive a new part) and the precise location and orientation of the part, and also be able to (1) sense the position and directional velocity of the human team-member, (2) understand (infer) the task that the person is engaged in, and (3) predict the next movements or actions of the person.

In addition to these components (human and co-robot awareness of the work task; mutual situation awareness of the other's state, activity, and future actions), processes of *team* situation awareness are also critical (Salas, Prince, Baker, & Shrestha, 1995). Team situation awareness implies a shared understanding of the situation and relevant elements in the task environment (e.g., in this case, progress through the manufacturing process, quality-related variables) developed through individual and team processes (e.g., individual information seeking, communication of information among team members; Salas et al., 1995). Team situation awareness is related to the more general concepts of transactive memory systems, shared mental models, and team cognition (Kozlowski, Grand, Baard, & Pearce, 2015; Morrow & Fiore, 2013). A transactive memory system includes shared knowledge of a team including both individuals' knowledge and the understanding of which team member knows what information (Wickens, Hollands, Banbury, & Parasuraman, 2013). Shared mental models include a shared understanding of the task or problem structure along with the knowledge and skills of other team members, and allow team members to work flexibly to adapt to changing work demands and to communicate more effectively (Morrow & Fiore, 2013).

Importantly, then, both team members should know what the other is aware of. That is, the human must understand what the robot knows about the work environment and the humans' movements and activities, and perhaps more critically, understand the limits of that awareness. This understanding is fundamentally tied to another theoretical lens—that of trust—discussed later in the chapter, and can be supported through explicit and transparent displays of not only what the co-robot is doing, but of the model that the co-robot has of the human partner. Team awareness is also reliant on successful communication, also discussed later in the chapter.

To date, although situation awareness in human-robot teams has been considered in a number of domains (Riley & Endsley, 2004; Riley & Endsley, 2005; Riley, Strater, Chappell, Connors, & Endsley, 2012, there has has been limited research specifically with regard to manufacturing teams involving co-robots. Gombolay, Bair, Huang, and Shah (2017) present one example, designed to address "one major gap in the robotics and human factors literature . . . the study of situational awareness wherein humans plan and execute a sequence of actions collaboratively within a human-robot team" (p. 600). They conducted research to understand how the situation awareness of the human team member changed across varying levels of robot autonomy in assigning tasks to human vs. robotic team members, in a situation where two humans and one robot collaborated on a physical assembly task, in part to understand how these systems should be designed to take into account the preferences of the human team member regarding task scheduling and workflow. They found that participant situation awareness of which team members were assigned and performed a task decreased when task assignment was done autonomously vs. by the human participant. However, this study did not address questions of situation awareness more broadly, in terms of mutual awareness of the other's actions in close proximity or shared awareness among the human and robot team member over aspects of the physical task. Additionally, it was not clear how (or if) the robot contribution to team situation awareness was assessed.

An important area of related research includes that of intent inferencing through the use of motion capture and kinematics. A current limitation of most co-robots that restricts their integration into a work cell is an inability to track human movement, estimate the future position, and react accordingly (Ivaldi, Fritzsche et al., 2017). Implementation of the co-robot requires knowledge of all tasks to be performed and detailed programming of the motion paths to be followed. True collaboration requires the robot to understand the human partner and to predict what the human will do next (Hayes & Scassellati, 2013; Ivaldi, Fritzsche et al., 2017). In order to achieve the first piece of this cycle, there is a need to track and model human dynamics. This includes both the body posture and the forces applied at the contact location, the motion and the effort. Most systems that monitor human motion near the robot use the information for safety purposes, allowing the robot speed to adjust when the operator is within predefined areas. Due to sensor costs and reliability concerns, this is traditionally limited to detecting the separation between the robot and the human and having the robot slow down when the distance is below a predefined threshold. Models of worker kinematics remain primarily in the research domain. Recent studies have focused on basic or isolated assembly tasks assigned between the human and co-robot with each agent working separately (Johannsmeier & Haddadin, 2017). For a screw-placing and sealing task, one study of human-aware motion planning showed ~5% faster task performance by the human, ~20% more concurrent motion, and reductions in idle time for both the human and robot (Lasota & Shah, 2015). An important result from this study was that the human-aware motion planning resulted in improved subjective evaluation of robot understanding, robot interference, and robot safety compared to the non-aware motion (Lasota & Shah, 2015). A human-co-robot movement chain is needed to model the shared task assignment for the human-co-robot team. Understanding this combined movement chain will support the goals of minimizing the interference between the human and co-robot and maximizing task performance. As indicated by Hayes and Scassellati (2013), anticipation of object placement and movement during task performance itself directly affects the possible actions and the available space for the collaborator (whether human or co-robot) on the team. Work cell design and task assignment can be affected by the incorporation of monitoring.

However, despite these advances, and in addition to solving the technical challenges related to sensing and motion capture, several important research questions exist when considering the problem of maintaining shared situation awareness in human-co-robot teams. One is the need to provide easy to understand displays or other signals that allow the human operator to develop and maintain an awareness over the co-robot actions. Current displays tend to be single indicators (e.g., red/green lights) that do not provide more detailed state information. Further discussion of human-co-robot communication is provided in the next section. More challenging, but critical to the creation of true team situation awareness, is the need to develop robust methods by which the co-robot can sense and appropriately understand human movement, infer activity, and predict future actions—without requiring explicit and continued input on the part of the human team member.

Communication

In a distributed cognitive system, development of a shared understanding—of the task situation and other team members' goals, activities, and level of shared awareness—depends on shared knowledge (including based on past experiences working together), shared awareness of the world in which the team is functioning, and successful communication about information which is required but not mutually known. For instance, human team members in a traditional manufacturing cell would, through past experience, understand task elements that are particularly challenging and also understand which team members have more or less expertise with certain tasks (and thus may require assistance). Because they are working together, at the same location, they share sensory input regarding, for instance, the operating state of machinery or characteristics of incoming parts. These shared experiences and shared understanding of the work environment contribute to the development of a "common ground," or a set of mutually shared knowledge, beliefs, and assumptions among communicative partners that support understanding and dialog (Clark & Brennan, 1991). Common ground—which encompasses background or fixed knowledge as well as aspects of the current situation and environment—is critical in facilitating, or in some cases reducing the need for, overt communication.

The need to explicitly communicate information and maintain shared situation awareness adds to the workload for distributed teams—team members must recognize the need for communication, act to communicate, check understanding, and effect repair of any communication failures. In human teams, maintenance of common ground and communication can be accomplished in subtle, sometimes implicit, ways that relieve some of this burden. For instance, facial expressions, gestures, or conversational tone can be used to direct attention, communicate confusion, or indicate the need for repair. Previous research on communication in complex work systems has documented the use of other, more implicit communication mechanisms. Co-workers may overhear conversations or observe others' actions and use that information as a cue to begin or inform their own activity (e.g., overhearing a phone conversation to initiate a station announcement in a mass transit control room; Heath & Luff, 1991; using an overheard confirmation of a ship's bearing in creating a positional plot; Hutchins, 1995) rather than requiring an explicit instruction or question. However, relying on auditory interfaces may be infeasible in manufacturing environments due to the ambient noise from machine operation.

Suchman's (1987) seminal work on situated action and human-machine communication revealed how the lack of common ground between human and machine team members could result in communicative error and task failure, even in a highly proceduralized task (making a photocopy). In her research, people were using a newly designed copier with advanced (for the time) capabilities, that included providing step by step directions to the human user and then inferring human intent based on limited sensing (e.g., whether a document had been placed on, or removed from, the glass). Inability to sense error and subsequent mischaracterization of the human user intent (i.e., what process step was being performed) led to an inability for the human-machine team to complete the task. This example, while outside of the manufacturing environment, is nonetheless telling: even for relatively straightforward, highly proceduralized tasks (e.g., as many manufacturing tasks are typically characterized), effective human-co-robot communication, including maintenance of common ground and detection and repair of communication failure, will be essential.

One thread of research related to human-co-robot communication has addressed the use of robot "facial expressions" or anthropomorphized display elements to convey information to human team members. For instance, a number of researchers are considering how robots might convey meaning or direct a teammate's attention through eye gaze behaviors in non-manufacturing environments (Mutlu, Yamaika, Janda, Ishiguro, & Hagita, 2009; Mwangi, Barakova, Diaz-Boladeras, Mollofre, & Roauterbert, 2018). Sauppe and Mutlu (2015) took an ethnographic approach to understanding human workers' social understanding and interaction with co-robots (specifically, the Baxter robot by Rethink Robotics) in three manufacturing work environments. They found that human co-workers were able to use signals from the robot (displayed "eye" movement and expressions) to infer the co-robot's status and intended next actions, and used other cues (e.g., operating sounds or lack of sounds) to identify or diagnose work problems. Operators also expressed a desire for additional communication capabilities, through speech and through strategic use of the robot's display screen (currently used to show the "eyes").

Other research has focused on the use of human gestures to communicate with or control co-robot actions. For instance, Tsarouchi, Matthaiakis, Makris, and Chryssolouris (2017) describe a human-co-robot hybrid assembly cell in which the human team member used gestures (sensed and recognized by the robots) to start, stop, or provide directional guidance to the robots. Importantly, however, the gestures were predetermined and part of a gesture "library" that could be understood by the co-robots, and thus represent an explicit means of communication, rather than an example of recognition of meaning in gestures which occur naturally during conversation or interpretation of human activity (ie. implicit communication). Related research regarding robot-produced gestures has studied how robots might use movement to indicate emotions (Li & Chignell, 2011; Novikova & Watts, 2015) and the interpretability of robot gestures by humans. For instance, Busch, Grizou, Lopes, and Stulp (2017) studied how a co-robot might "learn" to adapt their movements so they are more interpretable by people (specifically, allowing a person to interpret which button an industrial co-robot was going to press).

Important research and design questions related to human-co-robot communication therefore relate to the need to develop methods which support implicit as well as explicit communication, reduce workload related to explicit communication, support "natural" methods of communication and maintenance of common ground, and support effective detection and resolution of communicative problems. For instance, it should be possible for human team members to signal confusion or directives through voice and tone, using natural language. Co-robots should understand human movements and use that understanding to manage activity, without the need for explicit direction, and co-robots should use movements and gestures that are interpretable by their human teammates. Similar to recommendations regarding situation awareness, the co-robot's understanding of the work task, and its understanding of the intentions of the human team member, should be immediately and transparently apparent to the human team member.

Trust

Integral to the concept of high levels of human-co-robot collaboration is the notion of trust. Research on human trust in automated systems has a long history (Muir, 1994; Muri & Moray, 1996; Lee & Moray, 1994) and has leveraged models of trust developed in social science (e.g., Barber, 1983; Rempel, Holmes, & Zanna, 1985). A working definition of trust in the context of human-machine systems is provided by Lee and See (2004, p. 54): "the attitude that an agent will help achieve an individual's goals in a situation characterized by uncertainty and vulnerability."

Relevant findings from the body of research on trust in human-machine systems include the fact that there is at least some evidence which suggests that human concepts of trust are similar across trust in other people vs. trust in automated systems (Jian, Bisantz, & Drury, 2000), and that trust in systems grows through experience with successful operation, but also can be lost when systems do not operate as expected. Development and maintenance of trust is related to system reliability, and importantly, human expectation of and understanding related to the circumstances when automated systems should be expected to perform competently (e.g., concepts of robustness and understandability; Sheridan, 1988). Research in the context of automated decision aids suggests that human assessments of trust are higher when explicit information about the validity and reliability of system recommendations are provided (Seong & Bisantz, 2008).

Lee and See (2004) present a conceptual feedback model of factors which moderate behaviors of the intention to use, and actual reliance on, automated systems, including individual factors (i.e., perceived risk, self-confidence, predispositions to trust) and situation or system-related factors (e.g., workload, time constraints, effort to use). Feedback—that is, the experience of individuals relying on automation, and information gained about the automation through observation of its performance—impacts the development of trust as well as the "calibration" of trust. Calibration refers to appropriately understanding the contexts (temporal and functional) in which the automation should be trusted, or not. Importantly, in terms of design of co-robot displays or support for human-co-robot communication, Lee and See note that the "observability" of automation performance affects the development of trust. Related concepts include misuse stemming from overtrust (relying on a system even though it should not be trusted), disuse (abandoning a system in circumstances when it actually could aid performance Parasuraman & Riley, 1997), and complacency (failing to adequately monitor performance and therefore intervene when required, due to factors such as overtrust or workload; Parasuraman & Manzey, 2010). Also related, Meyer (2004) defines compliance as acting as directed by the system, while reliance is refraining from acting (e.g., exerting control) unless told explicitly otherwise. In a human-co-robot team, these behaviors would manifest (from the human perspective) as the human taking direction from the co-robot (compliance) or allowing the co-robot to work autonomously (reliance).

It is reasonable to assume that within a human-co-robot team, that the human team member will develop some attitude of trust related to the co-robot by observing its performance in the joint work environment, allowing the person to both rely on the co-robot to contribute to shared work tasks effectively, but also (and perhaps more importantly) to keep the person physically safe. Expectations related to co-robot competency can result from experience as well as directed training regarding expectations. Komatsu, Kurosawa, and Yamada (2012) describe an "adaptation gap" which arises when human expectations of a robot's behaviors (influenced by, for instance, anthropomorphic features) are not consistent with those behaviors. If a person's initial expectation of the robot exceeds the actual performance, attitudes towards use and acceptance would be negative. Similarly, unexpected behaviors due to co-robot faults (through mechanical or software failure) or programming changes (for instance, due to software updates) may negatively impact trust. In the case of human-co-robot teams, then, expectations set by training or factors such as the form or appearance of the co-robot regarding safety and competence should correspond to co-robot performance. For instance, Sauppe and Mutlu (2015) discuss how the co-robot's design features led to feelings of safety and comfort on the part of the human team members, but that such "increased sociality has the potential to create false expectations that may risk worker safety," requiring designers to match perceived and actual safety. Research in other complex systems has identified errors when interacting with automated systems in situations where otherwise highly competent systems unexpectedly lack specific capabilities: the human team member simply does not expect sophisticated automation to be unable to perform tasks that (from the human's perspective) are straightforward. For example, pilots were often "surprised" when sophisticated flight control systems did not protect them against inadvertently breaking a common altitude rule (Sarter & Woods, 1997).

There has also been some study of trust within the framework of human-robot interaction or human-co-robot teams. Sadrfaridpour, Saeidi, Burke, Madathil, and Wang (2016) created a mathematical, time-series-based measure of trust that was based primarily on the difference in task performance speeds between the human and co-robot team members. The work was premised on the assumption that humans would trust a co-robotic team member more if the co-robot adjusted its speeds based on human performance variability (e.g., slowing down due to muscle fatigue)—that is, if the co-robot adapted to the human team member's performance. Yagoda and Gillan (2012) developed a scale to measure trust in human-robot interactions which addressed factors beyond the robotic system to include team processes (e.g., communication, coordination, situation awareness), team configuration (other non-robotic team members), and aspects of task and the situational context. Charalambous, Fletcher, and Webb (2016) developed a scale to assess trust between humans and robots specifically in industrial settings (not specific to co-robots), finding that trust included factors related to assessments of the reliability of the robot gripper, the speed and fluidity with which the robot moved, safety in interactions, and comfort with features such as size.

Research challenges related to trust include helping the human team member appropriately calibrate trust regarding the co-robot's ability to perform the work task, but particularly with respect to safety—so that the human team member clearly understands the performance boundaries of the co-robot. The human team member must trust the co-robot in order to effectively collaborate in close physical proximity (a goal of these technologies). Co-robots are intended to reliably sense human motion or touch and avoid movements which release enough energy to be harmful to the human team member. Co-robots also could proactively warn human team members if the human is encroaching in an unsafe space (either related to the co-robot or to other aspects of the manufacturing environment). However, it is unlikely that the co-robot would behave in a way to "save" the human team member from some other unsafe act (e.g., deliberately disabling safety guarding) or unexpected danger (e.g., an out of control fork truck), in the way, for instance, that a human co-worker might act to warn or intervene.

Coordination of Action and Task Allocation in Distributed Teams

The coordination of action is fundamental to the functioning of even the most basic teams. Hutchins (1995) writes that "all divisions of labor, whether that labor is physical or cognitive, require distributed cognition to coordinate the activities of the participants" (p. 176) and uses as an example the coordination needs of two laborers working together to drive in a rail tie spike. At a micro-level, close coordination of action in both time and space requires team members to understand both the requirements for, and effects of, their own actions, as well as those of others. Team members must communicate (implicitly or explicitly), have knowledge about the task goals and results of actions, and share awareness of the current state of the work product and each others' actions. At a macro-level, negotiating the coordination of work tasks involves the allocation and often the dynamic re-allocation of tasks among team members.

Models of function allocation, particularly as related to level of automation, may provide some guidance. A number of frameworks for describing levels of automation have been proposed within the human factors literature, and include levels ranging from fully automated, to partial levels in which the human operator may or may not be involved in the choice or action and/or informed about action execution, to fully manual (Parasuraman, Sheridan, & Wickens, 2000). The framework proposed by Parasuraman et al. (2000) further breaks down the description of levels of automation across four task components—information acquisition, information analysis, decision selection, and action implementation—thus allowing a more differentiated model where task elements are automated to different degrees. These models can assist in characterizing the level of autonomy taken by the co-robot team member in selecting and completing tasks, or in contributing to a shared work task. For example, high levels of human-co-robot collaboration would generally require the co-robots to acquire information (through sensors), make inferences (regarding the state of the work task and human team member), and implement actions. However, the degree to which the robot determines the choice and timing of actions (vs. looking to the human team member for direction) could vary based on the work task, human preference, or the joint experience of the human and co-robot working together.

Research in task allocation across human and co-robot team members has been conducted from the perspective of optimizing task scheduling. For example, Tsarouchi et al. (2017) described the use of algorithmic, automated methods for assigning sequential tasks to humans or co-robots in hybrid assembly cells, based on parameters such as capability, task execution times, waiting time, and resource availability in order to maximize throughput and utilization rates. In the example cases, the optimized task allocation reduced completion time by 78% from the manual task time and brought down human resource utilization to one task. As suggested by the researchers, this would allow the human operator to work on other processes or across multiple cells. Gombolay et al. (2017) cite progress in computational methods that support task allocation and sequencing within human-robot teams, but still recognize that coordinating these team activities safely and efficiently is "a challenging computational problem."

Task coordination, however, is more than the assignment of tasks to team members. Gombolay et al. (2017) note that computationally-oriented research has not generally considered how to incorporate support for situation awareness of the human team member (regarding task assignments), if work and task assignments are made automatically, and specifically studied that issue (as described previously). Task coordination also requires adapting performance to the capabilities and needs of team members, which can change within a work shift, as team members change (e.g., interacting with individuals who might be left vs. right handed), or even within the performance of a single task. The research by Sadrfaridpour et al. (2016) considered how co-robots might need to adapt performance on one parameter (speed) in response to changes in the human team member (e.g., due to fatigue) and investigated various levels of autonomy in making that change, including in response to a request from the human team member, automatically by the co-robot, or a mixed approach, where either team member could request/initiate the speed change. Other research in co-robotics is considering coordination in tasks such as object handover between humans and co-robots, through methods which address the initial positioning of the robot and development of acceptable movement trajectories by taking into account the human team member's visual field and ease of reach (Sisbot, Marin-Urias, Broquere, Sidobre, & Alami, 2010; Broquere et al., 2014).

Importantly, however, these models and methods of function or task allocation do not seem poised to characterize some of the complex coordination which might be envisioned for the highest levels of human-co-robot collaboration. For instance, consider the actions of two human team members in a different, complex domain. Hutchins (1995) describes a set of highly skilled and coordinated actions among two naval personnel as they jointly held, slid, and rotated a plotting tool on a navigational chart, without any verbal exchange or explicit division of labor—"they simply created this coordination in the doing of the task" (p. 220). Likewise, two people might coordinate action in a joint assembly task (say, where one person is holding or positioning two parts, while a co-worker is using a tool to join the parts together). The first worker may subtly shift the part position to make it easier for the second to reach, or may change the position and force of his grip to counteract or assist in the actions of the co-worker using the tool. This implicit coordination, made possible through experience in performing the task, and in predicting trouble, might be supplemented with explicit requests for action and responses (e.g., "can you move that a bit so I can get a better view"). How (or even whether) co-robotic team members can be trained to achieve the levels of shared awareness, communication, and self-awareness of their own role in such task coordination is an open question.

INTEGRATING CO-ROBOTS ON TEAMS: KEY MANUFACTURING CHALLENGES

The theoretical frameworks described in this chapter reveal open questions regarding the nature of how people and co-robots can communicate and coordinate activities as members of a distributed manufacturing team. As cited previously, researchers in the broader area of human-robot interaction, particularly those focused in the area of social robotics, have taken on challenges related to creating more natural interactions between people and robots in non-manufacturing environments, including communication through natural language, use and interpretation of nonverbal communication, classification of human movements, developing awareness through sensing and interpretation of human movement, and developing environmental and situational models that support common communication ground. For instance, in creating the robot bartender JAMES, researchers are developing methods to use human positions, movements, gestures, eye gaze, and facial expressions to allow JAMES to interpret customer states (e.g., attending to a menu, interacting with other customers, attending to the bartender) to correctly determine which customers are waiting for service and manage interaction across multiple customers (Foster, Gashler, & Giuliani, 2017). There is significant interest and research in the area of robots that provide social connections, health monitoring, and task assistance for older adults (Robinson, MacDonald, & Broadbent, 2014), which for some tasks may require close physical coordination (e.g., for bathing or dressing assistance). Other research is addressing the common ground problem in "everyday" communication with a robot by developing methods for the robot to sense and model objects in the shared physical environment combined with what the human can see and might gesture to (Lemaignan et al., 2012). However, implementing human-co-robot teams in manufacturing environments requires additional, systems-level considerations, including impacts on safety, workplace, and task design.

Impacts on Operator Safety and Risk Assessment

One of the first, and continuing, concerns with the implementation of co-robots in manufacturing environments is the safety of the operator. Industrial robotics has traditionally involved robots moving at high speed, but separated from the operator by a physical barrier, such as a cage and with emergency stop buttons. With humans and co-robots sharing a workspace, these methods are no longer sufficient for protecting the operator from contact with the robot. The payloads and end-effector designs (which can include sharp, hot, or otherwise hazardous tips) required for manufacturing tasks pose additional safety risks beyond those of home or everyday social environments. Current planning for collaboration requires a detailed assessment of the tasks being performed and the risk presented by those tasks, regardless of whether one is designing a new work cell or modifying an existing cell into a collaborative work cell (Marvel, Falco et al., 2015). As part of this, research on adaptive automation for general human-robot interaction has considered the potential for human operator complacence and shown benefits to regular adjustment of expected operator interaction in encouraging maintained engagement (Parasuraman, Mouloua, & Hilburn, 1999; Parasuraman & Hancock, 2008; Scerbo, 2018).

Risk assessment for human-co-robot operation should be proactive and involve identifying hazards, mitigating risks, and documenting the process. Marvel et al. (2015) and the National Institute of Standards and Technology (NIST) have laid out a methodology for defining risks for human-co-robot collaborative tasks. While a thorough risk assessment process should still consider the severity and likelihood of the hazard, as well as the possibility of avoidance, a new, but equally critical, aspect for collaborative operation is separating the safety of the robot itself from the safety of the robot system (Marvel, Falco et al., 2015). As described earlier in the chapter, the robot system includes the robot and the accompanying end-effector and part attached at the end of the robot. The advancements in communication, situation awareness, and coordination between the human operator and the robot are necessary components for a robot to be safe. However, even if a specific robot has been deemed safe for one application, the addition of a cutting tool as the end-effector can change the hazards and risk level for the robot system, requiring further safeguards to be put in place. The end-effector and part are dependent on the task being performed and, therefore, the risk assessment should be performed at the task element level and consider the processes, tools, and environment involved (Marvel, Falco et al., 2015).

Addressing the identified risk for human-co-robot teams and tasks remains based on the hierarchy of safety controls (Barnett & Brickman, 1986). At the top, as the most effective control, is elimination of the risk through removing the hazard. This includes implementation of inherently safe design measures, such as choosing appropriate robot system components for the task. The next level is engineering control, preventing a hazard through programming and other means to separate the operator from the dangerous component. The third level is administrative control, changing the exposure of the operator to the hazard through work instructions, administrative restrictions, and other policy changes. Along with, and within, these levels, Michalos et al. (2015) described three main strategies for different safety types that are unique to human-co-robot interaction. Crash safety focuses on limiting the force exerted on the human operator in the case of a collision. Active safety employs a series of proximity and contact sensors to detect a possible collision, and stops the co-robot operation. Adaptive safety identifies necessary corrective actions in the planned co-robot motion path to avoid a collision and not stop the operation. It will be important for human team members to understand these safety strategies—both through training and during operation—to maintain appropriate levels of trust and coordinate activities with the co-robot.

The Robotic Industries Association (RIA) and the International Organization for Standardization (ISO) Technical Specifications provide guidance for system designers on hazard identification and risk assessment methods and necessary risk mitigation approaches for the design of safe collaborative workspaces (ISO/TS 15066, 2016). Of note, these recommendations and the proposed standard began with the starting point of existing standards for industrial robots, and thus focus more heavily on safety and risk assessment. Separate standards, including ISO 13482:2014, focus on safety requirements for personal care robots and robots used as medical devices (Harper & Virk, 2010). The safety standards provide a starting point for integrating collaborative robots, however they fail to consider other guidelines for safety and effective design, such as the use of gestures and social cues to support shared situational awareness and trust between the operator and the co-robot. Standards should be used in project planning, system requirements specification, and requirements verification (Harper & Virk, 2010), but with awareness of the limitations as a minimum level required to support operator safety.

Giuliani et al. (2010) described three design principles to support safe human-robot interaction that apply to standard industrial robots and to co-robots. These principles are robustness, fast reaction time, and context awareness (Giuliani, Lenz et al., 2010). The system must be robust to errors, compensating for and correcting incorrect inputs. It must also be able to respond quickly to sensor information gathered on the environment and the operator. Frustration and distrust of the co-robot can result from the perceived non-responsiveness of a slow co-robot. Advanced communication between the operator and co-robot can support the human's perception of the reaction speed. According to Giuliani et al. (2010), the robot system should be designed based on the specific context so that the robot is able to interpret ambiguous input information. These design principles affect the means of speech and vision processing used to support safe interaction. Templates and patterns for speech communication and gestures that the robot recognizes may be needed to allow for faster reaction from the co-robot.

Impacts on Work Cell Design and Ergonomics

Work cells include the space in which the human-co-robot team operates, along with the co-robot, the application software, machine vision system for detecting the operator, the tool and end-effectors, and the power source. Current methods for designing a safe work cell begin with performing a risk assessment, as described previously, and defining the collaborative workspace. The motion paths are then designed to be within the barriers and space constraints. Then the safety requirements for the robot are determined and the appropriate robot is selected. Robot selection depends on the tasks that need to be performed and scaled to the payload. Michalos et al. (2015) identified the following categories for requirements in work cell design: robot type (single vs. dual arm), robot payload and power generating ability, part characteristics (both geometry and weight), and the assembly and manufacturing processes that will be used. Typical co-robot payloads are 1 to 5 kg to support small parts assembly. Inappropriate selection, such as choosing a co-robot with larger capabilities, increases the difficulty of working collaboratively and safely, since the forces required to stop the co-robot increase (Zanchettin, Ceriani et al., 2016). Robot selection needs to include tests of the application in response to collision and error. Error can be caused by the human operator, a sensor malfunction, or an external source. In order to maximize productivity, the human and co-robot need to be able to work at similar speeds. Human operator task performance typically ranges from 500 mm/s to 1.5 m/s (Kamali, Moodie et al., 1982). Co-robots also differ in terms of their reach, repeatability, and accuracy of parts pick up and placement. The positioning requirements, accuracy requirements, and space and weight requirements need to be clearly defined prior to implementation. These requirements depend on a detailed task analysis and assessment of function allocation between the human and co-robot. Task factors and end effect selection do not only affect task performance. As described earlier, Charalambous et al. (2016) showed that trust of the co-robot by the operator was dependent on end-effector reliability, speed and smoothness of motion, and comfort of the operator with the size of the robot and end-effector.

Additional work cell ergonomics to consider include the means of communication between the operator and machines, the clarity of controls, and the intuitiveness of the interface available to the operator, in order to support shared awareness and task coordination, as described previously. In addition to understanding the general means of operation, the operator must know the contact needed to operate or stop the robot, or it should be built into the mode of collaborative operation. Many systems are also designed under ideal conditions with peak operator performance, but considerations of operator stress and fatigue should be factored into the work cell design.

As the introduction of a collaborative work cell involves a substantial investment, researchers emphasize the use of simulation to evaluate safe and effective design of the robot system. One digital human modeling tool proposed by Maurice et al. (2016) allows for comparison of different co-robots based on ergonomic outcomes for varying tasks. Robot systems can be compared on a series of 28 ergonomic indicators, from operator balance to awkward postures. Peternel, Tsagarakis, Caldwell, and Ajoudani (2018) have proposed a muscle fatigue model where the robot takes over the task once the human reaches a pre-defined fatigue threshold. As an advanced collaborative robotics research effort, the robot learns the task from the cooperation with the human so that it is prepared to take over once fatigue develops. Pearce et al. (2018) demonstrated an optimization approach for task scheduling of human-co-robot teams that minimizes both ergonomic risk and task duration. They showed that for tasks with poor hand/wrist posture (an indicator of ergonomic risk) and those that were repetitive, the method performed well for minimizing the idle time for both human and co-robot. However, tasks with high forces or that required high precision were not well addressed with the method. This approach applied the Strain Index as the means of identifying ergonomic risk, limiting implementation to only ergonomic risk to the distal upper extremity. Further work is needed on methods modeling the joint objectives of maximizing performance while minimizing negative human operator outcomes (whether they are physical or cognitive) for a broader range of tasks and conditions, and incorporating the workspace in the determination. A challenge is that these optimization models are typically run pre-implementation, and thus cannot dynamically adjust task allocation. Research is also needed on the integration of safety and human factors considerations of human-co-robot teams into design tools for manufacturing environments (such as Siemens Jack or CATIA), so that the collaborative work environment and the impact on the manufacturing process can be simulated (Michalos et al., 2015). It is also critical to define metrics to adequately assess human-co-robot team performance and determine what outcomes can be used to measure good interaction (Marvel et al., 2015).

AREAS FOR FUTURE RESEARCH

Co-robots have increased the flexibility of manufacturing process design, but, as described throughout this chapter, new challenges and research questions for human-co-robot teaming have been introduced. Here we highlight a few directions for future research into human-co-robot teaming in manufacturing environments. In this chapter, and in most current research and implementations, the human-co-robot team being considered has focused on one human and one co-robot working in a common workspace at the same time. Important questions remain regarding the challenges of asynchronous and remote collaboration, along with interaction as multiple co-robots and humans engage in teamwork. In larger teams there are concerns for the roles of each member and how individual and team situation awareness can be developed and maintained. As mentioned previously, critical to the creation of true team situation awareness is the need to develop robust methods by which the co-robot can sense and appropriately understand human movement, infer activity, and predict future actions—without requiring explicit and continued input on the part of the human team member. Accompanying such methods is the need to explore learning and adaptability of the co-robot. For example, many industrial co-robots are hand-guided during the training process to program initial movement plans based on expected task and worker positioning. This limits the flexibility of implementation unless new algorithms are programmed. With co-robots that could sense and predict human team member action, the system could learn from process data to minimize the likelihood of interference and potentially prevent errors from occurring.

Accompanying these technological questions, a number of important research questions exist related to human-co-robot communication and trust. One is the need to provide easy-to-understand displays that allow the human operator to maintain awareness over the co-robot actions. Even with the current typically pre-planned co-robot movements the displays tend to be basic indicators. As the teams become more complex or co-robot movement less predictable, investigating the best means of communicating detailed state information will become more critical. This includes the need for methods of implicit communication and the development of co-robot movements and gestures that can be interpreted by the human operator. Communication of co-robot intent and movement will also impact the calibration of trust for the co-robot's ability and safety.

CONCLUSION

The introduction of collaborative robots into manufacturing environments raises a number of important issues regarding the design of robots and the work environment. Clearly, as with any new manufacturing technology, manufacturing system designers are considering aspects of safety, quality, productivity, and ergonomic risk related to the use of co-robots. To the extent that co-robots are intended to take on true team member roles, however, critical design challenges remain. By examining these challenges, humans will be able to develop appropriately calibrated trust in their co-robot teammates and both human and co-robot team members will be able to develop a shared awareness of each other and the manufacturing task, mutually interpret each other's actions and intentions, and seamlessly and safely coordinate action.

REFERENCES

Barber, B. (1983). *The logic and limits of trust.* New Brunswick, NJ: Rutgers University Press.

Barnett, R. L., & Brickman, D. B. (1986). Safety hierarchy. *Journal of Safety Research, 17,* 49–55.

Broquere, X., Finzi, A., Mainprice, J., Rossi, S., Sidobre, D., & Staffa, M. (2014). An attentional approach to human-robot interactive manipulation. *International Journal of Social Robotics, 6,* 533–553.

Busch, B., Grizou, J., Lopes, M., & Stulp, F. (2017). Learning legible motion from human-robot interactions. *International Journal of Social Robotics, 9,* 765–779.

Charalambous, G., Fletcher, S., & Webb, P. (2015). Identifying the key organisational human factors for introducing human-robot collaboration in industry: an exploratory study. *The International Journal of Advanced Manufacturing Technology, 81,* 2143–2155.

Charalambous, G., Fletcher, S., & Webb, P. (2016). The development of a scale to evaluate trust in industrial human-robot collaboration. *International Journal of Social Robotics, 8,* 193–209.

Christiernin, L. (2017). How to describe interaction with a collaborative robot. In *HRI '17 companion, March 2017.* Vienna, Austria: ACM.

Chung, C. A. (1996). Human issues influencing the successful implementation of advanced manufacturing technology. *Journal of Engineering and Technology Management, 13*, 283–299.

Clark, H. H., & Brennan, S. E. (1991). Grounding in communication. In L. B. Resnick, J. M. Levine & J. S. D. Teasley (Eds.), *Perspectives on socially shared cognition*. Washington, DC: American Psychological Association.

Endsley, M. (1995). Towards a theory of situation awareness in dynamic systems. *Human Factors, 37*, 32–64.

Foster, M. E., Gashler, A., & Giuliani, M. (2017). Automatically classifying user engagement for dynamic multi-party human-robot interaction. *International Journal of Social Robotics, 9*, 659–674.

Giuliani, M., Lenz, C., Müller, T., Rickert, M., & Knoll, A. (2010). Design principles for safety in human-robot interaction. *International Journal of Social Robotics, 2*(3), 253–274.

Gombolay, M., Bair, A., Huang, C., & Shah, J. (2017). Computational design of mixed-initiative human-robot teaming that considers human factors: situational awareness, workload, and workflow preferences. *International Journal of Robotics Research, 36*, 597–617.

Harper, C., & Virk, G. (2010). Towards the development of international safety standards for human robot interaction. *International Journal of Social Robotics, 2*, 229–234.

Hayes, B., & Scassellati, B. (2013). Challenges in shared-environment human-robot collaboration. *Learning, 8*(9).

Heath, C., & Luff, P. (1991). Collaborative activity and technology design: Task coordination in London underground control rooms. In *Proceedings of the 2nd European conference on computer-support cooperative work, September, 1991.* Springer, Dordrecht.

Hutchins, E. (1995). *Cognition in the wild*. Cambridge, MA: MIT Press.

International Organization for Standardization. (2016). *ISO/TS 15066:2016*. Retrieved from https://www.iso.org/standard/62996.html

Ivaldi, S., Fritzsche, L., Babi⊠, J., Stulp, F., Damsgaard, M., Graimann, B., Bellusci, G., & Nori, F. (2017). *Anticipatory models of human movements and dynamics: The roadmap of the AnDy project*. Digital Human Models (DHM).

Jian, J. Y., Bisantz, A. M., & Drury, C. G. (2000). Foundations for an empirically determined scale of trust in automated systems. *International Journal of Cognitive Ergonomics, 4*(1), 53–71.

Johannsmeier, L., & Haddadin, S. (2017). A hierarchical human-robot interaction-planning framework for task allocation in collaborative industrial assembly processes. *IEEE Robotics and Automation Letters, 2*, 41–48.

Kamali, J., Moodie, C. L., & Salvendy, G. (1982). A framework for integrated assembly systems: Humans, automation and robots. *International Journal of Production Research, 20*, 431–448.

Komatsu, T., Kurosawa, R., & Yamada, S. (2012). How does the difference between users' expectations and perceptions about a robotic agent affect their behavior? *International Journal of Social Robotics, 3*, 109–116.

Kozlowski, S. W. J., Grand, J. A., Baard, S. K., & Pearce, M. (2015). Teams, teamwork, and team effectiveness: Implications for human systems integration. In D. Boehm-Davis, F. Durso, & J. D. Lee (Eds.), *APA handbook of human systems integration* (pp. 555–571). Washington, DC: American Psychological Association.

Lasota, P. A., & Shah, J. A. (2015). Analyzing the effects of human-aware motion planning on close-proximity human—robot collaboration. *Human Factors, 57*, 21–33.

Lee, J. D., & Moray, N. (1994). Trust, self-confidence, and operators' adaptation automation. *International Journal of Human Computer Studies*, *40*, 153–184.

Lee, J. D., & See, K. A. (2004). Trust in automation: Designing for appropriate reliance. *Human Factors*, *46*(1), 50–80.

Lemaingnan, S., Ros, R., Sisbot, E., Alami, R., & Beetz, M. (2012). Grounding the interaction: Anchoring situated discourse in everyday human-robot interaction. *International Journal of Social Robotics*, *4*, 181–199.

Li, J., & Chignell, M. (2011). Communication of emotion in social robots through simple head and arm movements. *International Journal of Social Robotics*, *3*(2), 125–142.

Marvel, J. A., Falco, J., & Marstio, I. (2015). Characterizing task-based human—robot collaboration safety in manufacturing. *IEEE Transactions on Systems, Man, and Cybernetics: Systems*, *45*, 260–275.

Maurice, P., Padois, V., Measson, Y., & Bidaud, P.(2016, June). A digital human tool for guiding the ergonomic design of collaborative robots. In *4th International Digital Human Modeling Symposium (DHM2016)*. Montreal, Canada.

Meyer, J. (2004). Conceptual issues in the study of dynamic hazard warnings. *Human Factors*, *46*(2), 196–204.

Michalos, G., Makris, S., Tsarouchi, P., Guasch, T., Kontovrakis, D., & Chryssolouris, G. (2015). Design considerations for safe human-robot collaborative workplaces. *Procedia CIrP*, *37*, 248–253.

Morrow, P., & Fiore, S. (2013). Team cognition: Coordination across individuals and machines. In J. D. Lee & A. Kirlik (Eds.), *The Oxford handbook of cognitive engineering* (pp. 200–215). New York: Oxford University Press.

Muir, B. M. (1994). Trust in automation: Part I. Theoretical issues in the study of trust and human intervention in automated systems. *Ergonomics*, *37*, 1905–1922.

Muri, B. M., & Moray, N. (1996). Trust in automation: Part II. Experimental studies of trust and human intervention in a process control simulation. *Ergonomics*, *39*(3), 429–460.

Mutlu, B., Yamaika, Y., Janda, T., Ishiguro, H., & Hagita, N. (2009). Nonverbal leakage in robots: Communication of intentions through seemingly unintentional behavior. In *Proceedings of HRI '09, March 11–13*. La Jolla, CA: ACM.

Muttray, A. et al. (2014). *Collaborative robots—Investigation of pain sensibility at the man-machine interface*. Institute for Occupational, Social and Environmental Medicine for the Johannes Gutenberg University of Mainz.

Mwangi, E., Barakova, E., Diaz-Boladeras, M., Mollofre, A., & Roauterbert, M. (2018). Directing attention through gaze hints improves task solving in human-humaniod interaction. *International Journal of Social Robotics*, *10*, 343–355.

Novikova, J., & Watts, L. (2015). Towards artificial emotions to assist social coordination in HRI. *International Journal of Social Robotics*, *7*, 77–88.

Parasuraman, R., & Hancock, P. A. (2008). Mitigating the adverse effects of workload, stress, and fatigue with adaptive automation. In P. A. Hancock & J. L. Szalma (Eds.), *Performance under stress* (pp. 61–74). Burlington, VT: Ashgate Publishing.

Parasuraman, R., & Manzey, D. (2010). Complacency and bias in human use of automation: An attentional integration. *Human Factors*, *52*, 381–410.

Parasuraman, R., Mouloua, M., & Hilburn, B. (1999). Adaptive aiding and adaptive task allocation enhance human-machine interaction. In *Automation technology and human performance: Current research and trends* (pp. 119–123). Mahwah, NJ: Lawrence Erlbaum Associates.

Parasuraman, R., & Riley, V. (1997). Humans and automation: Use, misuse, disuse, abuse. *Human Factors*, *39*(2), 230–253.

Parasuraman, R., Sheridan, T. B., & Wickens, C. D. (2000). A model for types and levels of human interaction with automation. *IEEE Transactions on Systems, Man, and Cybernetics-Part A: Systems and Humans*, *30*(3), 286–297.

Pearce, M., et al. (2018). Optimizing makespan and ergonomics in integrating collaborative robots into manufacturing processes. *IEEE Transactions on Automation Science and Engineering*, *15*(4), 1772–1784.

Peternel, L., Tsagarakis, N., Caldwell, D., & Ajoudani, A. (2018). Robot adaptation to human physical fatigue in human—robot co-manipulation. *Autonomous Robots*, 1–11.

Rempel, J. K., Holmes, J. G., & Zanna, M. P. (1985). Trust in close relationships. *Journal of Personality and Social Psychology*, *49*(1), 95–112.

Riley, J. M., & Endsley, M. R. (2004). The hunt for situation awareness: Human-robot interaction in search and rescue. In *Proceedings of the Human Factors and Ergonomics 2004 annual meeting*. Human Factors and Ergonomics Society. Los Angeles, CA: SAGE Publications.

Riley, J. M., & Endsley, M. R. (2005). Situation awareness in HRI with collaborating remotely piloted vehicles. In *Proceedings of the Human Factors and Ergonomics 2005 annual meeting*. Human Factors and Ergonomics Society. Los Angeles, CA: SAGE Publications.

Riley, J. M., Strater, L. D., Chappell, S. L., Connors, E. S., & Endsley, M. R. (2012). Situation awareness in human robot interaction: Challenges and user interface requirements. In F. Jenstsch & M. Barnes (Eds.), *Human-robot interactions in future military operations*. New York, NY: Routledge.

Robinson, H., MacDonald, B., & Broadbent, E. (2014). The role of healthcare robots for older people at home: A review. *International Journal of Social Robotics*, *6*, 575–591.

Robotics Industries Association. (2016). *RIA TR R15.606–2016 for robots & robotic devices—Collaborative robots*. Retrieved from https://www.robotics.org/robotics/technical-report-ria-tr-r15-606-2016-for-robots-and-robotic-devices-collaborative-robots

Sadrfaridpour, B., Saeidi, H., Burke, J., Madathil, K., & Wang, Y. (2016). Modeling and control of trust in human-robot collaborative manufacturing. In *Robust intelligence and trust in autonomous systems* (pp. 115–141). Boston, MA: Springer.

Salas, E., Prince, C., Baker, D. P., & Shrestha, L. (1995). Situation awareness in team performance: Implications for measurement and training. *Human Factors*, *37*(1), 123–126.

Sarter, N., & Woods, D. D. (1997). Team play with a powerful and independent agent: Operational experiences and automation surprises on the airbus A-320. *Human Factors*, *39*, 553–569.

Sauppe, A., & Mutlu, B. (2015). The social impact of a robot co-worker in industrial settings. In *Proceedings of CHI 2015* (pp. 3613–3622). Seoul, Korea: ACM.

Scerbo, M. W. (2018). Theoretical perspectives on adaptive automation. In *Automation and human performance* (pp. 57–84). Abingdon: Routledge.

Seong, Y., & Bisantz, A. M. (2008). The impact of cognitive feedback on judgment performance and trust with decision aids. *International Journal of Industrial Ergonomics*, 608–625.

Sheridan, T. B. (1988). Trustworthiness of command and control systems. In *IFAC Man-machine systems: Selected papers from the Third IFAC/IFIP/IEA/IFORS conference* (pp. 427–431). Oulu, Finland, 14–16 June 1988 .

Sisbot, E. A., Marin-Urias, L. M., Broquere, X., Sidobre, D., & Alami, R. (2010). Synthesizing robot motions adapted to human presence. *International Journal of Social Robotics*, *2*, 239–343.

Suchman, L. (1987). *Plans and situation action*. Cambridge: Cambridge University Press.

Tsarouchi, P., Matthaiakis, A., Makris, S., & Chryssolouris, G. (2017). On a human-robot collaboration in an assembly cell. *International Journal of Computer Integrated Manufacturing, 30*, 580–589.

Unhelkar, V. V., et al. (2018). Human-aware robotic assistant for collaborative assembly: Integrating human motion prediction with planning in time. *IEEE Robotics and Automation Letters, 3*, 2394–2401.

Wickens, C. D., Hollands, J. G., Banbury, S., & Parasuraman, R. (2013). *Engineering psychology and human performance* (4th ed.). New York: Routledge Publishing.

Yagoda, R., & Gillan, D. J. (2012). You want me to trust a ROBOT? The development of a human-robot interaction trust scale. *International Journal of Social Robotics, 4*, 235–248.

Zanchettin, A. M., et al. (2016). Safety in human-robot collaborative manufacturing environments: Metrics and control. *IEEE Transactions on Automation Science and Engineering, 13*, 882–893.

8 Subsidiary and Polycentric Control

Implications for Interface Design

John M. Flach and Kevin B. Bennett

CONTENTS

INTRODUCTION

> Each technology shift—manual to automated control to multi-layered networks—extends the range of potential control, and in doing so, the joint cognitive system that performs work in context changes as well. For the new joint cognitive system, one then asks the questions of Hollnagel's test:
>
> What does it mean to be 'in control'?
>
> How to amplify control within the new range of possibilities?
>
> (Woods & Branlat, 2010, p. 101)

Several years ago, we were observing a local regional disaster response exercise and we were surprised to discover that the police and firefighters had radios that could not be tuned to the same frequencies. At first, we were surprised and assumed that this was a consequence of antiquated equipment. However, a police commander overheard us commenting on this and he informed us that this was not a side effect of old technology. Rather, it was a deliberate choice. He told us that he did not want his police units to be getting instructions or commands from firefighters. The constraint on communications was designed to protect the lines of command against "noise" or "conflicting instructions" from other sub-organizations who were collaborating in the exercise.

In a subsequent analysis of the response to an actual regional emergency, we gained additional insight into how multiple organizations (e.g., police, fire, and medical) collaborate during a regional disaster (Flach, Steele-Johnson, Shalin, & Hamilton, 2014). We learned from our analysis that the emergency operation center (EOC) was not actually a command center, but rather, a communications center. That is, the primary function of the EOC was to make the collaborating sub-organizations aware of what resources were available or needed. However, decisions about how to access and use the resources were typically made within the collaborating agencies. Essentially, the function of the EOC was to facilitate situational awareness, so the individual agencies could accomplish their specialized functions. However, a necessary part of that situation awareness was information about the needs and resources of the cooperating agencies.

Subsidiarity in a Federation of Systems

Sage and Cuppan (2001) describe systems that involve cooperation among a collection of relatively autonomous agencies or subsystems to achieve a shared goal (such as we observed in the context of regional disasters) as a *federation of systems*. They make the case that a federalist style of organization can often achieve a good balance between global constraints (i.e., the need to work toward a common set of goals or within a shared value system) and the ability to adapt to local constraints and surprises that could not have been anticipated in advance. This balance is a critical factor determining the resilience of the organization in meeting the demands of complex work situations such as regional disasters. They note that an important consideration in the design of federations of systems is *subsidiarity*. Subsidiarity is the principle that authority for making a decision should *not* be reserved for a centralized command center, but rather the authority should be distributed and allocated to the lowest levels within an organization as possible (see also Rochlin, La Porte, & Roberts, 1987).

The experiences that McChrystal (2015) described with managing complex military operations echo and reinforce many of the observations made by Sage and Cuppan. The team of teams described by McChrystal is essentially a federation of systems. In particular, McChrystal emphasizes the importance of changing the culture to enable people closer to the field to take initiative and to make key decisions (i.e., subsidiarity).

Polycentric Control

Ostrom's (1999) analysis of how communities solve "tragedy of the commons"-type problems provides support for federalism and the principle of subsidiarity. The tragedy of the commons problem is created when communities depend on a common resource, creating a conflict between self-interest (maximizing your own harvest from that resource) and community interest (ensuring that the resource is not exhausted by over-harvesting so that it will be available for future generations). It had long been assumed that "tragedy" was inevitable, unless regulation was imposed by some centralized agent (e.g., a government agency). However, Ostrom (1999) found that attempts to solve these types of problems using centralized governance rarely succeeded. Communities that were able to solve the complex problem of balancing short-term and long-range consequences associated with sharing a pooled resource (i.e., the tragedy of the commons problem) often adopted a polycentric form of control.

A *polycentric control system* can be described as a system of systems (e.g., Sage & Cuppan, 2001). It is a layered collection of smaller organizations, each with their own local control structures (i.e., constraints), that cooperate toward achieving a long-term common goal or purpose. None of these smaller organizations has the capacity to address the full requisite variety associated with the long-term common purpose, but the solution that emerges as a result of the loose coupling across the local systems can achieve a satisfactory solution to the global problem.

The constructs of *polycentric control* and *subsidiarity* tend to challenge our usual notions of control that typically assume a centralized controller or authority. Based on her observations with communities who have avoided the tragedy of the commons, Ostrom (1999) has observed:

> The groups who have actually organized themselves are invisible to those who cannot imagine organization without rules and regulations imposed by a central authority.
>
> **(p. 496)**

In contrast with designed control systems (e.g., servomechanisms), federalist or polycentric control systems tend to be self-organizing systems. In such systems, stabilization around a satisfying state or goal tends to emerge from multi-level interactions, rather than being imposed in a hierarchical, top-down fashion. Thus, in addition to challenging our conceptions of "control," polycentric control systems challenge our conception of design, since polycentric control systems seem to have the capacity to reorganize themselves in order to adapt to dynamic changes in their work ecologies. In other words, polycentric control organizations tend to learn or to be self-organizing (e.g. Comfort, 2007; Gorman, Amazeen, & Cooke, 2010; Gorman, Dunbar, Grimm, & Gipson, 2017). A key question is, "What can be done to help ensure that a polycentric control system will learn and adapt in order to satisfy the functional goals of an organization in a dynamically changing ecology?"

Commander's Intent and Autonomy

The concept of *commander's intent* reflects one aspect of how the military is trying to balance the benefits of more distributed network forms of control with the challenge of maintaining coordinated action toward a common objective (Shattuck, 2000). On the one hand, the military would like to allow unit commanders the freedom to make the local decisions needed to adapt quickly to local constraints, but at the same time they realize that these decisions may have consequences that can be disturbances that require compensation by other units in order to achieve the common objective.

To provide more explicit details, the concept of commander's intent will be discussed in the context of Army tactical operations at the battalion level. A mission statement is received by the battalion from the higher-order echelon (i.e., brigade level). This mission statement is complete in the sense that "what" the commander wants to happen is specified. However, it should not be confused with a plan for the engagement. The guidance it contains, in the form of a commander's intent section, is fairly general in nature and surprisingly short (76 to 200 words; Klein, 1993). Klein (1994) analyzed commander's intent statements and found seven categories of information that were typically present: purpose (overall goals), objective (desired outcome), plans (expected general sequence of events), rationale (reasons behind the plan), key decisions (and contingencies), anti-goals (actions or outcomes to be avoided), and constraints (imposed limitations).

Note that not even the battalion commander is responsible for specifying the details of a mission plan. It is the responsibility of the lower-echelon commanders (in this case, the company commanders) to determine "how" the mission gets accomplished; they interpret commander's intent and generate the specific details that are required to fill in the mission plan. This plan is then communicated to lower-level units (i.e., companies and platoons) for execution through an operation order (OPORD).

The influence of the commander's intent does not stop at the planning stages of the tactical operation. This is made clear in the Army's definition of commander's intent:

> A clear, concise statement of what the force must do to succeed with respect to the enemy and the terrain and to the desired end state. It provides the link between the mission and the concept of operations by stating the key tasks that, along with the mission, are the basis for subordinates to exercise initiative when unanticipated opportunities arise or when the original concept of operations no longer applies.
>
> **(Army, 1997, pp. 1–34)**

Thus, commander's intent provides a unified basis for the intelligent adaptation of both plans and actions in response to the inherent variability and uncertainty of war. By specifying "what the force must do to succeed" instead of "what the force must do" (i.e., specific mission plans), the commander's intent provides lower-level commanders with the flexibility to adapt dynamically to changing circumstances in the battlefield.

Note, however, that this increased flexibility comes with a cost. Specifically, the benefits of adaptivity are offset by the costs of coordination. Subordinate units (in this case, company commanders) cannot act in a vacuum. Changes in mission plans must be accompanied by a high degree of communication and coordination with other subordinate units if they are to succeed. Once again, this process is guided by commander's intent (Headquarters Department of the Army, 2003):

> 6–52. Plans are not static; commanders adjust them based on new information. During preparation, enemies are also acting and the friendly situation is evolving: Assumptions prove true or false. Reconnaissance confirms or denies enemy actions

> and dispositions. The status of friendly units changes. As these and other aspects of the situation change, commanders determine whether the new information invalidates the plan, requires adjustments to the plan, or validates the plan with no further changes. They adjust the plan or prepare a new one, if necessary. **When deciding whether and how to change the plan, commanders balance the loss of synchronization and coordination caused by a change against the problems produced by executing a plan that no longer fits reality. The higher commander's intent guides their decision making.**

In summary, the construct of commander's intent suggests that the solution involves a loose coupling between the central command and the field units. The central command specifies the overall objectives, but the unit commanders supported by network communication capabilities are given the authority to adapt as required to the local constraints and the responsibility to communicate with other units to facilitate coordination toward achieving the common objectives. The result is that while central command specifies the operational goals, they are not involved in micro-managing the actions of specific units in adapting to local disturbances.

As with the EOC described previously, a critical factor enabling increased autonomy of unit military officers involves communications. A network of communications among the unit officers is essential to support mutual adjustments. In colloquial terms, in order to achieve coordinated action toward a common goal, the right hand has to know what the left hand is doing. When communications are sparse, mutual adjustment is not possible, and thus coordination will be more dependent on following predetermined plans or following instructions from a higher-level agency that has access to the "big picture."

In line with the construct of commander's intent is the military's desire for increasingly autonomous technologies. In this context, "autonomy" reflects a technology's ability to adapt intelligently to situations or circumstances without direct supervision from a human. This is consistent with the principle of subsidiarity. However, there is always the question of "How much autonomy, or how much discretion do you want machines or junior officers to have?" With complete discretion or autonomy there is a significant risk that the subsystems will be working at cross purposes and the common organizational goals will not be achieved (e.g., as in the tragedy of the commons).

Requisite Variety and the Degrees of Freedom Problem

Two important considerations in the design of all control systems are Ashby's (1956) Law of Requisite Variety and Bernstein's (1967) Degrees of Freedom Problem. The *Law of Requisite Variety* says that a controller must be at least as complex as the system it is controlling. In other words, it takes variety to destroy (control) variety. On one hand, this suggests that teams will generally have greater capacity for control than individuals and that diverse teams will have greater capacity than homogeneous teams. This principle points to the obvious benefits of collaboration. Teams and organizations will generally be able to solve problems (e.g., a regional disaster) that individuals cannot accomplish alone.

On the other hand, the *Degrees of Freedom Problem* calls attention to the costs of collaboration. Keeping multiple units or individuals moving toward a common goal can be extremely demanding. The more independent units involved (the more degrees of freedom) the more difficult it will be to coordinate activities in order to avoid having units work at cross purposes and to make progress toward a common goal (e.g., satisfying the commander's intent).

In the context of motor control, Bernstein suggested that skilled athletes manage this tradeoff through assembling coordinative structures. Bernstein noted that a large number of degrees of freedom in the motor system allows people to do many different motor activities (e.g., drive, chip, or putt a golf ball). However, he noted that to be skilled at any of these specific activities it was essential to reduce the number of degrees of freedom that needed to be coordinated for any particular action. Thus, he suggested that skilled athletes are able to function as different specialized "machines" (or coordinative structures). Each machine involves locking out many of the degrees of freedom, so that only a few dimensions have to be adjusted in real time.

We believe the intuitions of Bernstein can be generalized to the design of high-functioning teams and organizations. In addressing the challenge of subsidiarity and answering questions like how low in the organization you can push authority or how much discretion or autonomy you can give to individuals, we are essentially designing coordinative structures. By moving authority to subsystems within an organization and constraining communications within and across subsystems, we are essentially locking out some degrees of freedom in favor of others. Thus, in addition to distributing authority, a critical factor in managing complexity will be appropriately constraining the communication structure, that is, designing a network of communications that (1) meets the need for quick action to adapt to local disturbances, (2) meets the need for coordination with other agents toward a common goal, and (3) does not overwhelm or overload any of the agents with too much information.

We believe that constructs such as polycentric control, subsidiarity, commander's intent, and self-organization have important implications for how we design interfaces. The remainder of this chapter will offer hypotheses about how to design interfaces to support polycentric control.

INTERFACE DESIGN

Beginning in the 1980s the technologies for computer graphics have been improving dramatically, and this has given designers many new options for how to display and manipulate information on computer screens (e.g. Greenberg, 1991; Hutchins, Hollan, & Norman, 1986; Shneiderman, Plaisant, Cohen, Jacobs, & Elmqvist, 2018; Woods, 1984; Woods, 1991). This has created the opportunity to consolidate information that in the past had been distributed over many mechanical instruments into graphical displays with configural geometries that are capable of highlighting relations among the individual variables that are difficult to "see" in the single-sensor-single-indicator configurations. The design of these configural displays was inspired by research in Gestalt psychology (e.g., Wertheimer, 1959) that demonstrated the significant impact that a change of representation can have on the difficulty of problem solving. Rasmussen and Vicente (1989) introduced ecological interface design (EID) as a framework for how to identify functional constraints at different levels of abstraction and how to represent these constraints as configural display geometries that supported supervisory control and fault diagnosis (e.g., see Bennett & Flach, 2011; or Borst, Flach, & Ellerbroek, 2015 for more recent reviews of the EID approach).

However, Segal (1994) has observed an unintended potential negative side effect of integrating and optimizing interfaces to support individual problem solving. He suggested that as a consequence of centralizing information into a single graphical representation, nonverbal information that had been available to support team coordination may be lost. For example, in a multi-operator control room with information distributed over walls of analog meters, it was possible to make strong inferences about what other operators where thinking about by seeing where they were looking and by observing what they were doing. However, in a control room with all the operators sitting at computer terminals with configural graphic displays, there is little or no information to allow anyone to "see" what other people on the team are thinking about. Segal suggested that as displays became more consolidated, it was necessary for people to compensate for the loss of nonverbal cues for coordination with increased levels of verbal communication.

Segal's observations support the importance of understanding cognition at the team level. His work illustrates that a focus on optimizing individual cognition without consideration for the impact on coordination can actually undermine performance at the organization level. Thus, in the remainder of this chapter we will consider the challenge of designing interfaces that include coordinating with other agents as part of the problem to be solved.

Common Operating Picture

A number of years ago, we had the opportunity to observe within a command center during a complex military exercise. The command center was about the size of a high school gymnasium and it was filled with 50 to 60 workstations organized around various operational cells involved in the exercise (e.g., offensive ops, defensive ops, refueling, special ops, time-sensitive targeting). Four large screen displays were mounted on one wall of the command center at a level that could be seen from any of the workstations in the room. These screens were intended to supplement the information available at the individual workstations in order to provide a common operating picture to support coordination among the cooperating cells.

Unfortunately, little thought had been given about what to put on these screens. An early attempt to show everything was scrapped immediately because it was readily apparent that the displays were so cluttered with information that it was impossible to pick up anything useful. The solution that was implemented during the exercise was to mirror the interfaces of selected individual cells. At the end of the exercise, it was found that people did not find these wall displays to be very useful.

For the most part, the primary tools for coordination during the exercise were chat rooms and the "sneaker net." At most workstations, one monitor tended to be dedicated to chat rooms; individuals routinely monitored from three to nine of these chat rooms as one means to coordinate within and across cells. The term "sneaker net" was used to describe coordination events that occurred when subsets of people gathered for face-to-face discussions to resolve particularly difficult issues. Later research by Courtice (2015) indicates that chat and voice communication systems tend to play complementary roles for coordination. Chat tends to be used to protect the voice channel, which tends to be reserved for particularly difficult problems. Thus, as with Segal's (1994) observations, verbal channels appear to provide very critical resources for coordination.

Semantic Mapping and Work Analysis

The motivation behind the use of the wall displays during the command and control exercise (i.e., to help establish a *common operating picture*) was well grounded (e.g., Wolbers & Boersma, 2013). The problem was in the execution. Specifically, there was a failure to do a systematic semantic or work analysis. This activity is necessary to explore alternative ways to parse and organize information on the wall displays so that there was a direct mapping to the semantic properties of the work. As Bennett and Flach (1992) have noted, a key to designing representations to support problem solving is semantic mapping: the degree to which meaningful properties of the problem are mapped to perceptually salient properties of the representation.

For the EID approach, Rasmussen's (1986) abstraction-decomposition space has been suggested as a useful framework for decomposing work to preserve semantic relations (see Flach & Voorhorst, 2016; Vicente, 1999 for more detailed descriptions of this space). As illustrated in Figure 8.1, Rasmussen (1986) found that during problem solving, experts tend to navigate along the diagonal of this space, such that concepts at higher levels of abstraction tended to provide a context for organizing information and guiding attention at lower levels of abstraction. Rasmussen (1986) notes that

> At the lower levels of functional abstraction, elements in the description match the component configuration of the physical implementation. When moving from one level to the next higher level, the change in system properties represented is not merely removal of details of information on the physical or material properties. More fundamentally, information is added on higher-level principles governing the cofunction of the various elements identified at the lower level. . . . Change of level of abstraction involves a shift in concepts and structure for representation as well as a change in information suitable to characterize the state of the function or operation at the various levels of abstraction. Thus, a decision maker will ask different questions regarding the state of the physical system, depending on the level of abstraction that is most suited to formulate the actual control task.
>
> **(pp. 19–20)**

Rasmussen found that more abstract-aggregate properties tended to provide the general functional motivations (why) for the work and tended to set broad constraints bounding the range of potential affordances (i.e., possibilities for action that exist in the work domain; Gibson, 1979). He found that more concrete-particular details

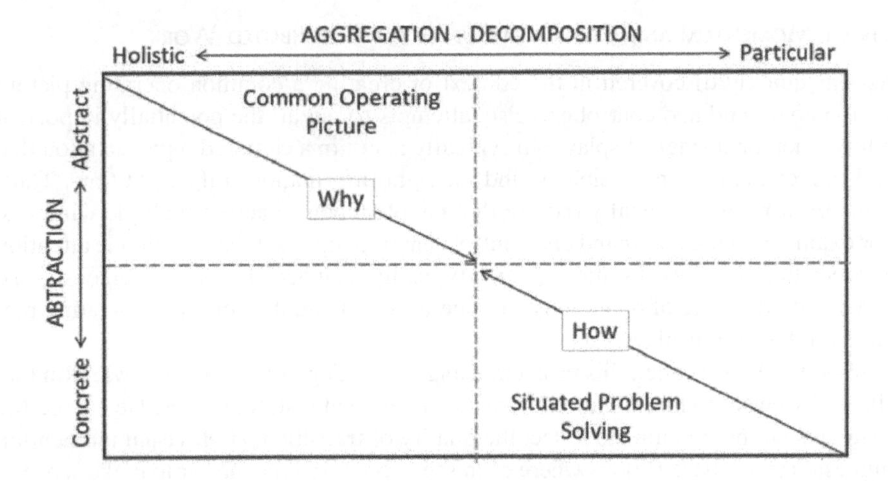

FIGURE 8.1 Rasmussen has discovered that during problem solving, experts tend to navigate along the diagonal of this abstraction-decomposition space, such that abstract-aggregate properties tend to direct attention to specific details.

Note: This insight has implications for distributing information across multiple displays to support coordination in a polycentric control system.

became important for choosing the specific affordances to fit specific situations. In line with Rasmussen's comments, we suggest that in designing wall displays to provide a common operating picture to support coordination it is important to consider the upper left quadrant of the abstraction-decomposition space (see Figure 8.1). Information in these displays should be organized around functional purposes, goals, and the higher-level principles that govern the co-functioning of the elements, while leaving the lower-level details to individual workstations supporting situated problem solving.

Thus, for example, the wall displays for the command and control center might have been designed to reflect key functional distinctions: offensive objectives, defensive threats, and logistics. The fourth display might have been designed for re-planning or adapting to events not addressed in the initial plans (e.g., emergent, time-sensitive targets or rescue operations). Note that this parsing of the displays corresponds roughly to the functional distinctions associated with the various cells in the command center. However, the point of the wall displays is to provide information about the different functional goals that are to be shared across the cells, rather than the details needed for operations within a cell. Thus, the information in the wall displays should be in terms of holistic/aggregate aspects associated with the distinct functional goals, rather than the particular operational details. For example, the function of the offensive wall display would be to let other cells know about the general progress toward satisfying the objectives (i.e., the commander's intent) and to help them judge whether additional resources are needed or available. In particular, once an objective has been satisfied, the assets allocated to that objective might be potential resources for utilization to execute a target of opportunity identified in the time-sensitive targeting cell.

Visual Momentum and Computer-Supported Distributed Work

As was quickly discovered in the context of creating a common operating picture for the command and control exercise, attempts to put all the potentially important information on a single display will typically result in a cluttered representation that makes it difficult or impossible to find the right information at the right time. Thus, complex work will typically require that people navigate across multiple windows. For example, in the command and control center, people will have to divide attention between their local workstations (which typically included at least two screens—one focused on the state of operations and one for chat) and the common operating pictures on the wall displays.

Woods (1984) framed this problem using the analogy of editing a movie film that allowed viewers to make smooth transitions between distinct scenes. He coined the term "visual momentum" to reflect the quality of transition. High visual momentum suggested smooth transitions where each scene prepared the viewer to make sense of the next scene and low visual momentum suggested difficult transitions that required cognitive effort to make sense of the new scene. He suggested a number of design solutions that might contribute to creating smooth transitions, including fixed format data replacement, long shots (zoom and pan), landmarks, and overlap (see Bennett & Flach, 2012 for recent review).

These techniques should also be considered as promising techniques for supporting coordination within a polycentric control system. While each of the different agencies cooperating within a polycentric control organization might require different information details reflecting their specific functions, there is also a need to create the common ground for coordination. Techniques such as including common long shots or common landmarks across the specialized interfaces might help satisfy this design goal. For example, in the military command and control example, the key aspects of the battle space (e.g., political boundaries, significant geographical boundaries such as coastlines, and significant locations of friendly and enemy bases) might provide key landmarks that would support smooth transitions between displays that may be very different in detail (e.g., satellite, topological, and political maps).

In addition to Wood's suggestions for improving visual momentum, there is an extensive literature on computer-supported collaborative work (CSCW) (e.g., Greenberg, 1991; Gutwin & Greenberg, 2001; Gutwin, Greenberg, & Roseman, 1996; Lyytinen & Ngwenyama, 1992; Pratt, Reddy, McDonald, Tarczy-Hornoch, & Gennari, 2004; Schmidt, 1991). In part motivated by Segal's (1994) concerns about the information that can be lost when computer workstations replace physical workspaces and in part motivated by the opportunities that computers offer for supplementing other sources of information, the CSCW literature is an important source for ideas about how interfaces can be designed to support collaboration. This literature considers enhancing performance in shared workspaces (where people are co-located in the same physical space) and in distributed workspaces (where people are working from distributed locations).

The CSCW literature explores how various technologies (e.g., video conferencing) can bridge the distances in a distributed control system to help people cooperate to solve complex problems. While much of the CSCW research is framed in terms of the various technologies, a key theme in the CSCW literature that aligns with the theme of this chapter is the need for a thorough work analysis to understand the functional semantics associated with collaborative work. As suggested by Segal's (1994) observations, the ultimate goal is often to recover important information for coordination that gets lost when people do not share the same space. Gutwin and Greenberg (2001) introduce the construct of workspace awareness to emphasize the significance of this information for achieving shared objectives:

> when we work with others in a physical shared space, we know who we are working with, what they are doing, and where they are working.
>
> **(p. 6)**

SUMMARY AND CONCLUSIONS

Hopefully, this chapter has provided some general direction for people seeking to design interfaces to support collaboration in polycentric control systems. The chapter does not go into great detail, because the details will vary with each work organization and each domain. Rather, the goal was to provide some general guidelines. The most important message of this chapter is the need for any design approach to be guided by a deep understanding of the organizational-level dynamics. To this end, the first half of the chapter focused on new ways to think about "control." This reflects a view that being in control is fundamental for accomplishing work.

The case is made that to work successfully in an increasingly networked world, adaptive, polycentric forms of control (i.e., self-organization) are required, and that an important principle for designing polycentric control systems is subsidiarity. That is, distributing the authority for making decisions from a centralized control center to local agencies (or individuals) who have the benefit of tight coupling with local situations. However, this must be balanced by communications to support mutual adjustment so that local adaptations do not interfere with progress toward a common goal.

It is important to recognize that while networks allow great flexibility for adaptation to changing situational demands, they also facilitate the propagation of noise that can lead to interference and overload. Thus, it will often be necessary to introduce constraints to enable stable control. The critical design issue is to align the constraints with the functional objectives—so that the organization functions as a variety of interconnected, simple machines that together span the situated demands of the work. This involves a layered control system where an outer, adaptive layer is responsible for picking the right machine for each situation. Each simple machine is responsible for its functional niche and for pushing the information to other machines relative to shared resources and goals.

These general principles of control then provide a framework for pursuing a detailed work analysis to uncover the semantics of a particular work domain, in which collaboration is an essential part of that semantics. In other words, some awareness of what our collaborators are doing and what resources they offer (or need) is essential to avoid working at cross purposes and enabling mutual adjustments toward achieving common goals. This awareness is typically referred to as "common ground."

With respect to the details of work domain analysis, there are interesting parallels between Rasmussen's description of work in an abstraction-decomposition space and the layering and separation of authority/responsibility in a polycentric control system. It is suggested that these parallels might be useful in deciding how to construct interfaces to provide a common operating picture, without overwhelming people with details.

Additionally, we suggest that Wood's construct of visual momentum and the associated recommendations for display techniques that foster smooth transitions across interfaces might also be considered as techniques for building common ground. This suggests that coordination across time and coordination across agencies may benefit from similar design interventions.

Due to advances in automation, computation, and networking the nature of cognitive work is becoming more complex. Thus, the requisite variety exceeds the capacity of most individuals, making collaboration among multiple humans and automatons essential for success. An important way in which designers can help to shape these collaborations toward productive thinking is in the design of interfaces. These interfaces must be designed to simultaneously provide the situated details needed for adaptation to local disturbances and surprises and provide the common ground to support teamwork (mutual adjustment) toward a shared goal. While there are many suggestions in the CSCW literature for interventions to support collaborations, implementation of these suggestions must be guided by a deep understanding of the nature of the cognitive work to be supported.

REFERENCES

Ashby, W. R. (1956). *Introduction to cybernetics.* London: Chapman & Hall.

Bennett, K. B., & Flach, J. M. (1992). Graphical displays: Implications for divided attention, focused attention, and problem solving. *Human Factors, 34*(5), 513–533.

Bennett, K. B., & Flach, J. M. (2011). *Display and interface design: Subtle science, exact art.* Boca Raton, FL: CRC Press.

Bennett, K. B., & Flach, J. M. (2012). Visual momentum redux. *International Journal of Human-Computer Studies, 70*, 399–414.

Bernstein, N. (1967). *Coordination and regulation of movements.* New York: Pergamon.

Borst, C., Flach, J. M., & Ellerbroek, J. (2015). Beyond ecological interface design: Lessons from concerns and misconceptions. *IEEE: Transactions on Systems, Man, and Cybernetics, 45*(2), 164–175.

Comfort, L. K. (2007, December). Crisis management in hindsight: Cognition, communication, coordination, and control. *Public Administration Review*, 189–197.

Courtice, A. M. (2015). *Chat communication in a command and control environment: How does it help.* Unpublished dissertation, Wright State University, Dayton, OH.

Flach, J. M., Steele-Johnson, D., Shalin, V. L., Hamilton, G. C. (2014). Coordination and control in emergency response. In A. Badiru & L. Racz (Eds.), *Handbook of emergency response: Human factors and systems engineering approach* (pp. 533–548). Boca Raton, FL: CRC Press.

Flach, J. M., & Voorhorst, F. A. (2016). *What matters? Putting common sense to work.* Dayton, OH: Wright State University Library.

Gibson, J. J. (1979). *The ecological approach to visual perception.* Boston, MA: Houghton Mifflin.

Gorman, J. C., Amazeen, P. G., & Cooke, N. J (2010). Team coordination dynamics. *Nonlinear Dynamics, Psychology, and Life Sciences, 14*, 265–289.

Gorman, J. C., Dunbar, T. A., Grimm, D., & Gipson, C. L. (2017). Understanding and modeling teams as dynamical systems. *Frontiers in Psychology, 8*, 1053.

Greenberg, S. (1991). *Computer-supported cooperative work and groupware*. London: Academic Press.

Gutwin, C., & Greenberg, S. (2001). *The importance of awareness for team cognition in distributed collaboration*. Report 2001–696–19. Dept Computer Science, University of Calgary, Alberta, Canada.

Gutwin, C., Greenberg, S., & Roseman, M. (1996). Workspace awareness in real-time distributed groupware: Framework, widgets, and evaluation. In R. J. Sasse, A. Cunningham & R. Winder (Eds.), *People and computers XI (proceedings of the HCT96)* (pp. 281–298). Amsterdam: Springer-Verlag.

Headquarters Department of the Army. (2003). *Field manual 6–0, Mission command: Command and control of army forces*. Washington, DC: Author.

Hutchins, E. L., Hollan, J. D., & Norman, D. A. (1986). Direct manipulation interfaces. In D. A. Norman & S. W. Draper (Eds.), *User centered system design* (pp. 87–124). Hillsdale, NJ: Erlbaum.

Klein, G. A. (1993). *Characteristics of commander's intent statements* (Tech. Rep., Contracts MDA903-90-C-0032 and MDA903-92-C-0098). Arlington, VA: U.S. Army Research Institute for the Behavioral and Social Sciences.

Klein, G. A. (1994). A script for the commander's intent statement. In A. H. Levis & I. S. Levis (Eds.), *Science of command and control: Part III. Coping with change* (pp. 75–86). Fairfax, VA: AFCEA International Press.

Lyytinen, K. J., & Ngwenyama, O. K. (1992). What does computer support for collaborative work mean? A structurational analysis of computer supported cooperative work. *Accounting Management and Information Technologies, 2*(1), 19–37.

McChrystal, S. (2015). *Team of teams*. New York: Penquin.

Ostrom, E. (1999). Coping with tragedies of the commons. *Annual Review of Political Science, 2*, 493–535.

Pratt, W., Reddy, M. C., McDonald, D. W., Tarczy-Hornoch, P., & Gennari, J. H. (2004). Incorporating ideas from computer-supported cooperative work. *Journal of Biomedical Informatics, 37*, 128–137.

Rasmussen, J. (1986). *Information processing and human-machine interaction*. New York: North Holland.

Rasmussen, J., & Vicente, K. J. (1989). Coping with human errors through system design: Implications for ecological interface design. *International Journal of Man-Machine Studies, 31*, 517–534.

Rochlin, G. I., La Porte, T. R., & Roberts, K. H. (1987). The self-designing high-reliability organization: Aircraft carrier flight operations at sea. *Naval War College Review, 40*(4), 76–92.

Sage, A. P., & Cuppan, C. D. (2001). On the systems engineering and management of systems of systems and federations of systems. *Information-Knowledge-Systems Management, 2*(4), 245–325.

Schmidt, K. (1991). Riding the tiger, or computer supported cooperative work. In *Proceedings of the second European conference on computer-supported cooperative work, September 1–16*. Amsterdam, The Netherlands: Kluwer Academic Publishers.

Segal, L. (1994). *Effects of checklist interface on non-verbal crew communications*. NASA Contractor Report 177639. Moffett Field, CA: Ames Research Center.

Shattuck, L. (2000). Communicating intent and imparting presence. *Military Review, 80*(2), 66–72.

Shneiderman, B., Plaisant, C., Cohen, M., Jacobs, S. Elmqvist, N (2018). *Designing the user interface*. Essex, UK: Pearson Education Limited.

U.S. Army. (1997). *Field manual (FM) No. 101–5–1: Operational terms and graphics*. Washington, DC: Headquarters Department of the Army.

Vicente, K. J. (1999). *Cognitive work analysis*. Mahwah, NJ: Erlbaum.

Wertheimer, M. (1959). *Productive thinking*. New York: Harper & Row.

Wolbers, J., & Boersma, K. (2013). The common operating picture as collective sensemaking *Journal of Contingencies and Crisis Management, 21*(4), 186–199.

Woods, D. D. (1984). Visual momentum: A concept to improve the cognitive coupling of person and computer. *International Journal of Man-Machine Studies, 21*, 229–244.

Woods, D. D. (1991). The cognitive engineering of problem representations. In G. R. S. Weir & J. L. Alty (Eds.), *Human-computer interaction and complex systems* (pp. 169–188). London, UK: Academic Press.

Woods, D. D., & Branlat, M. (2010). Hollnagel's test: Being "in control" of highly interdependent multi-layered networked systems. *Cognition, Technology, and Work, 12*, 95–101.

9 The Cognitive Wingman

Considerations for Trust, Humanness, and Ethics When Developing and Applying AI Systems

Michael D. Coovert, Matthew S. Arbogast, and Ewart J. de Visser

CONTENTS

TRUST IN HUMAN-MACHINE SYSTEMS

As we begin our consideration of trust and what it means in the context of human-machine teams, especially when AI enters the mix, it is important to "begin at the beginning" and define our core construct, namely trust. There are many definitions of trust, but arguably two have become more dominant than others (Lewicki & Brinsfield, 2017). The first is offered by Rousseau, Sitkin, Burt, and Camerer (1998, p. 395), whose focus is on how trust makes us vulnerable to one another. For these authors, trust is the "intention to accept vulnerability based upon positive expectations of the intentions or behavior of another." Acceptance of vulnerability seems a reasonable description of trust in human-machine systems as well, for if an individual utilizes a technology such as a self-driving car or the autopilot in a plane, one is certainly exposing oneself to the consequences of the action of the technology. A second definition is proposed by McAllister (1995, p. 25), "the extent to which a person is confident in, and willing to act on the basis of, the words, actions and decisions of another." Here the central focus is on the relationships between individuals or entities. In our context, one of the entities is a technology. As Lewicki and Brinsfield (2017) point out, research around these two perspectives of trust illustrate it is multidimensional, having three factors: cognitive, affective, and behavioral. The cognitive factor relates to those specific expectations and beliefs relating to the other; affective refers to the emotional connection with the trusted entity; and behavioral refers to those actions taken by an individual in a trusting relationship. In addition to these three dimensions of trust, it should be noted there is emerging evidence for a neurological basis. Examination via fMRI reveals that trust and mistrust involve different functional areas of the brain (Krueger et al., 2007) with certain neurochemicals (oxytocin) also being involved (Zak, 2017). While we do not delve into it here, this does impact those interested in brain-computer interfaces for monitoring levels of trust and the use of neurochemicals for manipulating susceptibility to trusting.

Trust is one of those ideas that everyone knows about and experiences. Yet, in terms of operationalizing the construct, there are many boundary conditions to consider; fortunately, there is also much agreement. For extensive reviews, visit Lewicki and Brinsfield (2017) for interpersonal trust and Lee and See (2004) for trust in automation. Bringing this vast literature into focus for the purpose of our chapter, we highlight a few topics central to trust in technology before moving to consider humanness and ethical issues in Artificial Intelligence (AI) systems.

Trust in Technology

When thinking of trust, one often considers such notions as competence, reliability, and benevolence (McAllister, 1995). This is certainly true for trust in interpersonal relationships, and we believe it extends to as trust in technology as well. We now describe some exemplar studies central to trust in AI systems.

Measuring Trust and Modeling Its Growth

To understand trust and how it changes in the context of either interpersonal relationships or technology, we must be able to measure it and then model it. Coovert, Miller, and Bennett (2017) demonstrated the utility of latent growth models and latent change score models for both measurement and modeling purposes. They examined the growth of the two types of trust (cognitive and affective) found in McAllister's (1995) framework. Coovert et al. demonstrate trust in teammates develops first with cognitive trust and once that is established, affective trust begins to grow. Utilizing bivariate coupling with other constructs of interest (e.g., satisfaction with teammates, team performance) the growth of trust within a dynamical system can be modeled along with other parameters of interest and importance.

Facilitating Trust

A system, especially one with advanced technology, will not be used until it is trusted. Given this truism, several research programs have been developed to determine how to facilitate the development of trust. Recently, Lyons, Ho et al. (2016) focused on the factors associated with the use of a controlled flight into terrain (CFIT) avoidance system in aircraft. The purpose of the system is to take over control of flying the aircraft should the pilot be flying too close to hazards (or in the extreme become incapacitated, as can happen when a plane is performing maneuvers exerting high G-loads on the pilot). Lyons et al. expressed three key takeaways from their study. The first is the algorithms (used to take control) must be nuisance-free, in that you do not want to warn a pilot about an impending crash too early. This finding is concomitant with the false alarm rate described later in the chapter. As it turns out, there are individual differences associated with preferences for the timing of the warning. Since individual differences are involved, this is a parameter that could be set by each pilot. For example: If *airspeed* is greater than A, *closure-rate* greater than or equal to B, and *hazard-typ*e is C then display CFIT warning type *Bravo*. Pilots are individually allowed to set values for *airspeed*, *closure-rate*, *hazard-type*, and CFIT *warning type*. Since each individual configures the system according to their own preference, this leads to increased propensity to trust.

A second finding also relates to individual differences. When aircraft control is taken by the system it should fly the evasive maneuvers consistent with a pilot's preference for flying those maneuvers. Trust is personal, so if you want the pilot to trust the automation, it needs to take over in a fashion that is consistent with the pilot's mental model of preferred actions. The third finding relates to building trust through training. By using training to assess, demonstrate, and verify the reliability of the systems, one builds trust in the system.

Consider the findings from the above research: (1) understanding false alarms and why they occur, (2) grasping a system's parameters and how they can be tailored for individual preference, and (3) evaluating fit with one's mental model. Is there a common factor underlying these findings? One can argue *transparency* is such a factor. The idea of transparency has been examined by several researchers working in the area of human-robot interaction (HRI). In laying a groundwork for transparency, Kahn et al. (2012) demonstrate that humans hold robots accountable, at least more accountable than they hold other inanimate objects such as a toaster that burns bread or the ice maker that fails to make ice in a freezer. Development of effective HRI interactions occur by having individuals understanding the robot's ability, intent, and situational constraints (Coovert et al., 2014). Aspects of these ideas were also posed by Kim and Hinds (2006), who suggest transparency should increase as automation accelerates along a dimension anchored by complete autonomy.

Furthermore, Burke, Coovert, Murphy, Riley, and Rodgers (2006) and Lyons (2013) argue people need to understand robots and the robots need to have an "understanding" of people. Lyons suggests the robot-to-human transparency should take place on several levels, with separate models for intention, task, analytical, and environment. He proceeds to provide the following description and elaboration of the models. An *intentional model* conveys the "why" of the system. This helps place the human's understanding of the system in the correct context. It presents the design, purpose, and intent of the system. In addition to "why," it also provides "how" the robot performs the actions. Similar to Asimov's (1942) three laws of robotics, it provides an understanding of the robot's moral philosophy of interacting with humans (we discuss moral and ethics issues below). A *task model* includes information relating to the robot's cognitive goals at a given time, information relating to progress toward those goals, information signifying an awareness of the robot's capabilities, and awareness of errors. An *analytical model* communicates the underlying analytical principles used by the robot to make decisions. For instance, knowing that a particular robot fuses information from satellite imagery and ground sensors in determining where potential emergency zones are located could be useful if the human knew the ground sensor network had been compromised (Lyons, 2013). This is similar to the algorithm transparency principal describes in the CFIT research. Finally, the *environmental model* provides the human with information about what the robot is experiencing.

The idea of transparency is not limited to embodied complex technologies such as robots. Organizations are now grappling with when and how to reuse complex computer code. The ability to reuse code has several advantages, two of which are of immediate benefit to an organization. First, as a pragmatic issue, the code is already paid for. The second factor relates more directly to our concern: code that is in use is *trusted* code. Alarcon et al. (2017) discuss issues relative to the reuse of code, and much of the argument could be viewed through the lens of transparency. The authors used interviews and cognitive task analysis to identify issues associated with the reuse of code. Several factors emerged. Firstly, the reputation of the code considering its source, reviews, and number of users—each of these is an indicator of trust. Secondly, transparency—factors such as organization, style, architecture, and comments each influences how understandable the code is. Thirdly, performance of the code reflected in metrics for efficiency, resiliency, relevant functionality, flexibility, and being error free. Fourthly, environmental factors play a role. These include customer needs and requirements, organizational and resource constraints, and consequences of failure. Finally, Alarcon et al. believe individual differences such as propensity to trust and personality will also impact the decision to reuse code.

In summary, several models play a role in the transparency of a technological system and the human's ability to gain trust with the complex technology, be it a robot, AI system, or other type of advanced technology. To further facilitate the building of trust, one might consider providing a technology the ability to assess the human's state (fatigued, overloaded, frustrated, afraid, angry). An example is the case for monitoring a driver's state to allow technological intervention during instances of high fatigue (May & Baldwin, 2009).

If we agree transparency is important, then we must determine how best to introduce it into the system. One obvious place is via the interface, but a second is arguably more impactful and that is through extensive training. By utilizing effectual training, the operator gains an understanding of the models employed by the system, for example with the robot (intentional, task, analytical, environmental) or reuse of code (reputation, transparency, performance, environmental). We do not claim that all technological systems need to have these models; some may need more and others may function quite well with fewer, but for many technologies those systems seem to be an effective blend. Whatever the case, in order to build trust and maximize human-machine synergy it is necessary for the technological systems to be transparent in terms of their goals and operations.

Is the meaning and use of transparency in technological systems transparent? That is to say, there may be many types of transparency and it is important to identify those which are most effective with different technological systems. Lyons, Koltai et al. (2016) empirically examined two different types of transparency in terms of trusting and using an emergency landing planner (ELP) for commercial airliners. The traditional system developed by NASA provides standard information to pilots without explanation of the benefits (or costs) of alternative courses of action. These authors looked at transparency of explanation as impacting acceptance of a recommendation made by the ELP. They used a within-subjects design and considered two types of explanatory systems. The control condition used the standard ELP (e.g., weather, runway characteristics, and terrain). The first transparency system utilized risk-based transparency (referred to as the *value* condition). It gave all the information as the base condition but also provided a probability of success (e.g., 39% chance the flight crew will be able to successfully complete the approach and landing under current conditions). The second condition was based on logic-based transparency, whereby a statement was included to provide the logic behind the risk statement (e.g., the runway is unacceptable because the crosswind is too high for a safe landing). Results demonstrated trust in the system was the highest in the logic-transparency condition, followed by the risk-based transparency, and lowest in the control condition. These findings clearly support the premise that advanced technology will not be used without trust in the system. Here the output of a system is not accepted on a prima facia basis. Rather, trust in the system is facilitated through an explanation of how the technology arrived at the recommendation. This explanation gets at the heart of the models employed by the technology and makes everything transparent to the user.

False Alarms—Inhibitors of Trust

There are many factors that impact trust among workers in an organization. For example, Galford and Drapeau (2003) describe how false feedback, inconsistent messaging, and inconsistent standards are among those factors negatively impacting trust in the workplace. Since many, if not most of these systems will be associated with workplace tasks, we must consider issues of workplace and interpersonal trust, as well as trust in automation. If a system is to be trusted it must, at a minimum, provide reliable information. Yet many systems have multiple technological components. For example, consider the cockpit of a modern aircraft. There are many individual gauges and displays (e.g., attitude, airspeed, altitude) providing information regarding the status of the aircraft. The same can be said for the dash in the driver's area of a car, providing information regarding the status of the car (e.g., speed, traction, oil pressure). A question arises as to how individuals establish a level of trust in such situations; is it based on each individual component sensor or display (e.g., separate trust levels for attitude, airspeed, and altitude) or is one level of trust established for the system as a whole? Gees-Blair, Rice, and Schwark (2013) examined this issue through the manipulation of false alarm rates. Their work indicates individuals form one impression of the system as a whole and increased levels of not just faulty, but also false alarms in one device/display will impact trust in the entire system. Thus, an individual may be unwilling to utilize a technology (e.g., fly the aircraft, ride in a self-driving car) because impressions of the unreliability of one component spreads (contaminates) the perception of the reliability of the system as a whole.

Once trust in a technology degrades or ceases due to real failures, degradation of performance, or false alarms, if the technology is to be used again we must deal with the process of trust repair. Just as humans must engage in a progression whereby trust between two individuals must be rebuilt due to some action or inaction on the part of another, so too must we consider trust repair in the context of human-machine teams. Technology will give false alarms, make poor recommendations, take inappropriate actions, and will fail to act when appropriate. Trust between humans and machines will need to be repaired and we should know how best to proceed. Much work needs to be done in this nascent area and those interested should see de Visser, Pak, and Shaw (2018) who argue we should use models of how relationships between two humans are repaired to guide us in the process of how human-technology trust repair should be fashioned. Another useful model is presented by Kim, Dirks, and Cooper (2009) who offer an interesting bilateral model of trust repair that accounts for perceptions and characteristics associated with both the human trustor and human trustee. Their approach to understanding trust negotiation efforts and trust repair methods are worth additional study and application. Furthermore, Kim et al. suggest that unfixable flaws in the person (e.g. their character) are hardest to repair. This suggests that the moral fiber and ethical decision making of an agent is the greatest concern; integrity and benevolence is key to maintaining a healthy trust-based relationship and has implications for understanding morality and trust in cognitive agents (more on this later in the chapter).

We began this chapter introducing definitions of trust containing facets of vulnerability, action, cognition, and affect, among others. Trusting someone means putting ourselves at risk; it also means allowing someone or something to act on our behalf. For this to effectively occur the literature is clear that various aspects of the technology must be transparent to the user. This transparency will come in many types depending on the technology, and may be best presented to the user via the technology's interface or through training and experience with the technology. Just like humans, however, the technology will not be eternally omniscient and act flawlessly. Therefore, trust between the human and the technology will need repair. We need to develop theories and strategies associated with trust repair in human-machine systems. Of course, none of this can be done without reliable and valid means to measure and model trust.

We now move to examine humanness, as much of what we know about technology and how it will be trusted, accepted, and utilized will depend on the degree to which it is anthropomorphized. Following our discussion of humanness, we move to consider technology, especially AI technology as our "cognitive wingman."

TRUST, HUMANNESS, AND MORALITY

DAVE BOWMAN: Open the pod bay doors, HAL.
HAL: I'm sorry, Dave. I'm afraid I can't do that.

—2001: A Space Odyssey

Movies such as *The Terminator*, *I Robot*, *The Matrix*, and *2001: A Space Odyssey* have portrayed robots and non-human intelligent agents with ill intent. The basic premise of these films is that machine intelligence and awareness inevitably result in conflict and adversarial relationships with humans, often with lethal results for the human (de Visser et al., 2018; Snyder & Mcneese, 1987; Fraser, Hipel, Kilgour, McNeese & Snyder, 1989). More recent science fiction, however, paints a subtler relationship between humans and intelligent machines. In *Moon*, the robot GERTY is assigned to monitor the health and schedule of a human employee but hides communications and directives from the company. In *Robot & Frank*, a health provider robot prescribes meals and exercises that are (initially) rejected by Frank. In *Her*, the operating system Samantha exists to serve as a personal assistant, but instead pursues her own hidden goals and initiatives. And in the recent *I Am Mother*, a robot raises a child in a world absent of other people and hides the true extent of her master plan. In sum, the popular conception of non-human intelligent agents has shifted somewhat towards increased nuance, away from overt aggression and toward subtle conflicts of interest.

Recent work by futurists and philosophers have proposed the inevitable rise of the super intelligent machine, that is, a machine that creates its own machines which are more intelligent than any human being currently alive. These futurists say the super intelligent agent may be the last invention we ever have to create but warn that it may be mankind's undoing. Others are more optimistic and propose that super or artificial intelligence will generally be a force for good to be embraced by everyone.

The possible creation of super intelligent machines poses an interesting problem for the scientific community, because current frameworks on machine intelligence primarily address automation, adaptive automation (Byrne & Parasuraman, 1996; Scerbo, 1996, 2008; Feigh, Dorneich, & Hayes, 2012), and more recently autonomy (Kaber, 2018a). The frameworks and theories are not equipped to deal with intelligence that is not based on human intelligence. Thus, such intelligent machines will have an impact on the types of frameworks we use to normally classify automation and technology (Kaber, 2018b; McNeese, 1986).

The Cognitive Agent Spectrum

Although the stories from the movies in the previous section are mainly for entertainment purposes, some of these imagined realities may not be far from resembling our own. As non-human agents have become better able to mimic human intelligence (or affect certain elements of human intelligence—the increasing social "charm" of Apple's Siri or Amazon's Alexa, for example), research is needed that accounts for the various roles that will be fulfilled by these agents. Specifically, greater understanding is required of the implications of automated aids with "ulterior motives," that is, those whose stated goal (e.g., informing an unbiased purchasing decision) might differ from the goal the user perceives the agent as having (e.g., guiding them toward a particular purchase). Improvements in technology and design will allow human-machine interactions to be more complex and nuanced. As the quality of these interactions becomes closer to human-human interactions, a greater understanding of the role of machine "humanness" should be pursued. Although the human-automation research literature has traditionally assumed a dichotomy between humans and automation (Madhavan & Wiegmann, 2007; Nass, Moon, Fogg, Reeves, & Dryer, 1995; Nass, Steuer, & Tauber, 1994), the increasing social and analytic capabilities of automation may soon necessitate the consideration of automation "humanness" along a continuum. There have been classifications of machines along a functional spectrum (Kaber & Endsley, 2004; Parasuraman, Sheridan, & Wickens, 2000) but we propose a *cognitive-agent spectrum*. Such a classification would arrange machines along a continuum of cognitive agent (cognitive independence): the degree to which a machine is perceived as initiating, executing, and controlling its own actions. Variables that determine the degree of independence ascribed to a machine may include complex factors such as the level and sophistication of interaction capabilities as well as simpler qualities like appearance (e.g., lines of code vs. an expressive computer avatar). With these novel agents varying on humanness, it becomes important to investigate how we perceive and trust these agents.

The Uncanny Valley

The Uncanny Valley, a theoretical effect proposed by Mori (1970), suggests that as human attributes are emulated by machine agents with increasing fidelity, subjective feelings of liking will increase until that emulation becomes near perfect, at which

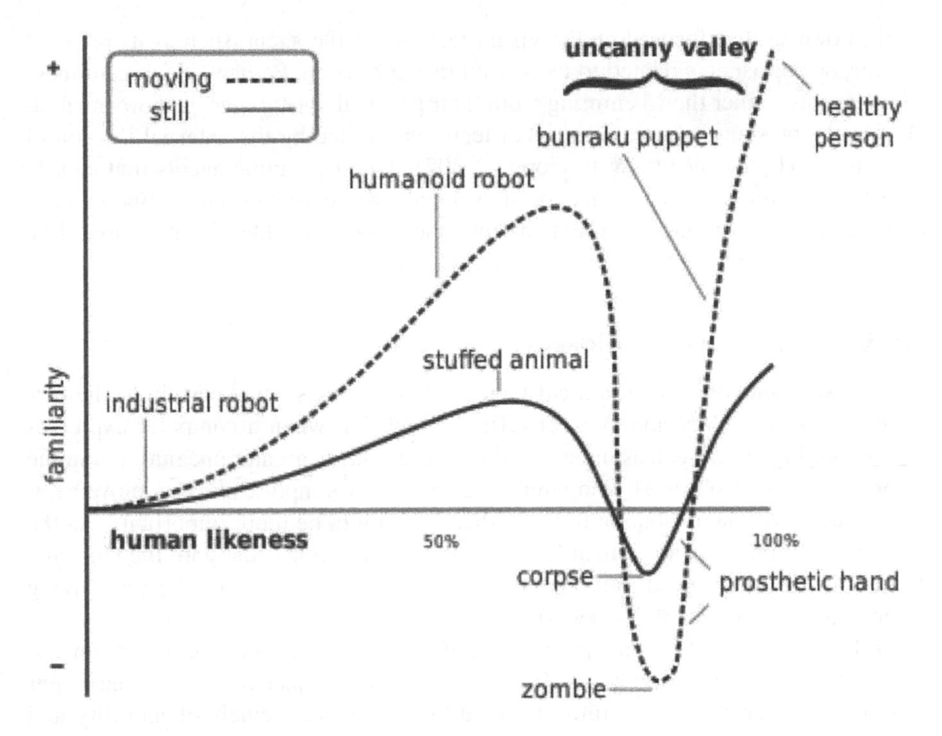

FIGURE 9.1 The Uncanny Valley.

Source: Figure adapted from MacDorman and Kageki (2012).

point the agents will evoke feelings of discomfort instead (see Figure 9.1). The sudden decrease in liking as robotic agents approach human appearance is the "valley." Since this theory was proposed, it has received a great amount of attention in the human-robot interaction community; if valid, the Uncanny Valley has direct implications for the design of the humanoid robots.

The empirical support for this theory remains mixed, however. Some research has elicited effects consistent with the valley using still images of humans and machines morphed together (MacDorman & Ishiguro, 2006) or using images of abnormal features (Seyama & Nagayama, 2007) or images of monkey faces (Steckenfinger & Ghazanfar, 2009) or human faces in decision support systems (Wellens, 1993). Research using videos of robots-in-motion, however, has typically failed to elicit an Uncanny Valley effect, which is somewhat surprising given that motion is theoretically supposed to augment the effect (C.-C. Ho, MacDorman, & Pramono, 2008; C. Ho & MacDorman, 2010; K.F. MacDorman, 2005; Saygin, Chaminade, Ishiguro, Driver, & Frith, 2011). Rather than causing a valley, the effect for increasing fidelity of humanoid movement seems to be a monotonic increase in perceived humanness and familiarity (Thompson, Trafton, & McKnight, 2011).

Research on the Uncanny Valley has thus far failed to provide a consistent account for the role of humanness in perception of humanoid agents. The majority of this work, however, has focused on the visual features of the agent, such as its physical appearance or aspects related to its motion through space. Perhaps the uncanniness of the agents, rather than stemming from their physical appearance, is more attributable to the presence of certain internal features implied by the external illusion of humanness (Epley, Waytz, & Cacioppo, 2007). That is, robotic agents that closely approximate human appearance or movement (while remaining distinctly non-human) cause the human observer to assume the robot is capable of other human-like functions, like goal-directed behavior or emotion.

The Uncanny Valley and Morality

Recent work suggests that a critical factor for uncanniness may pertain to the perception of mind. Gray and Wegner (2012) found that when a computer expresses experiential features such as hunger and fear, it produces greater uncanniness in the mind of a human observer compared to a robot that simply expresses movement-related agency. The attribution of mind therefore might be more important than the perception of physical movement for eliciting an effect consistent with the Uncanny Valley. Indeed, perceived internal experience (consciousness) may be the driving force behind the Uncanny Valley effect.

If the perception of internal processes and preferences induces feelings of unease among human observers, it seems likely that robotic agents expressing a particular moral preference would be similarly disturbing. A recent review of morality and theory of mind suggests that moral preference and perceptions of intent are difficult to disentangle (Gray, Young, & Waytz, 2012). Although no research has yet examined user perceptions of a morally dubious robot, some research has begun to explore the idea of moral accountability for robotic aids. Using a monetary incentive, Kahn et al. (2012) found that participants held a robot partner morally accountable when the robot failed to perform in the incentivized task. Ratings after the experiment showed that most participants rated the robot as morally accountable for the prizes that it caused the participant to lose. The fact that participants held the robotic aid morally accountable begs some comparisons to how they might have acted with a human partner. Although participants rated the robot's social and experiential features as being significantly less prominent than those of a human, the prominence of these features was rated as being greater than those of a simple vending machine. Although Kahn et al. (2012) did not test this premise, we expect that participants would not have held a vending machine morally accountable for its malfunctions. Thus, the potential distinction between various non-human machines in terms of their perceived moral responsibility hints at the presence of a continuum. What remains unclear, however, are the factors that might influence the degree to which individuals or teams of people perceive various cognitive agents as being morally culpable.

Morality and Trust in Cognitive Agents

Tightly linked to the idea of morality in agents is the matter of trust. Trust is informed by how we expect an entity might act towards us in the future. In fact, perceptions of moral character have been shown to directly impact levels of trust. Delgado, Frank, and Phelps (2005) found that participants who read written biographies describing a morally "bad" partner were less likely to rely on that partner in a subsequent economic trust game. In contrast, levels of trust for morally "good" partners were significantly elevated over the "bad" and "neutral" ones. Participants also shared more of their winnings with the good partner compared to the bad partner. This experiment clearly supports the idea that moral character influences trust and subsequent cooperation-related decision making.

Research has not yet examined how moral character influences the perception of humanness for non-human agents, and how these perceptions might affect trust and ratings of liking. Some research suggests that human-human interactions are directly comparable to human-machine interactions, and that differences in trusting are only quantitatively (rather than qualitatively) different (Nass et al., 1994). Consequently, this research proposes that automation can elicit social interactions that are similar in quality to the interactions occurring between humans. In support of this view, a recent study did not find any differences in trustworthiness of a medical robot tasked with taking a blood pressure reading when it had a silver face, a human face, or no face at all (Broadbent et al., 2013). In direct opposition to this media equation hypothesis, some researchers propose that trust between humans is *qualitatively* different than trust between humans and machines (de Visser et al., 2012; Dzindolet, Peterson, & Pomranky, 2003; Lee & See, 2004; Madhavan & Wiegmann, 2007). This automation theory has also found empirical support: when playing the ultimatum game, people are much more willing to accept unfair offers from a computer than they are from a human, perhaps due to a stronger emotional response towards the humans (Sanfey, Rilling, Aronson, Nystrom, & Cohen, 2003). It would therefore be useful to examine this issue further incorporating a theoretical framework such as McAllister's (1995) since his incorporates two different types of trust, cognitive and affective. That theoretical perspective would predict differential level of cognitive and affective trust toward the targets, human vs. technology.

Morality with Super Intelligent Agents

Researchers have proposed the possible advent of the ultra-intelligent machine (Good, 1965), a machine that can far surpass all the intellectual activities of any human, no matter how clever. If this machine could design even smarter machines, then this is the last invention man would have to make provided the machine is docile enough to tell us how to keep it under control (Good, 1965). Recent reports have speculated about when mankind may invent such a machine with dates ranging from early to late 21st century (Bostrom & Yudkowsky, 2014; Kurzweil, 2005). This also raises the possibility that we would attribute a mind to such a machine and that we would also imbue it with a sense of morality, perhaps even a superior morality to our own. It is doubtful people will accept the morality of a machine if it means that they have to give up control. Every movie about the future, at least in the West, plays with the theme of fear as a consequence of losing control to a machine. Current debates focusing on the use and policy of autonomous weapons raise a lot of emotions for this reason. Yet, it is not unthinkable that a machine, in some cases, could act with superior morality compared to man. A lot of our personal morality is gut-based, whereas society's morality is rule and law based. There is an opportunity here to create machines that have human values encoded into them and act as moral agents of humanity. In this sense, we are in the unique position of defining our morality more precisely and having it more faithfully executed by machines. Design guidelines on how to implement these kinds of moral rules should be a primary focus of those in the human factors and broader scientific community. If successful guidelines are created, and healthy human-machine trust relationships are established, we can then fully apply artificial intelligence capabilities to enhance distributed cognitions and effectiveness among teams.

AI: OUR "COGNITIVE WINGMAN"

As outlined in the beginning of this chapter, the importance of cognitive-based trust (McAllister, 1995) and trustworthiness is simply indisputable (Coovert et al., 2017) and it is difficult to overstate the positive influence these constructs have on effectiveness of distributed teams. Although we consider trust and trustworthiness baseline requirements for performance, exploring elements beyond these constructs is vital to harnessing the full potential of each team. Given the current state of technology, there is a unique opportunity to embrace AI and the projected benefits may very well outweigh the risks. Leveraging AI as a cognitive "wingman" and maximizing trust among human-machine teammates can lead to numerous applied solutions relevant to distributed team cognition.

We do recognize that there is a plethora of dire, fear-mongering predictions regarding the use of AI, such as those proposed by Elon Musk and Stephen Hawking (Clifford, 2018). Although we agree that warnings against blindly trusting intelligent agents should not be ignored, strategies built on risk avoidance can prove costly and hinder efforts to maintain a competitive advantage. Unquestionably, there may be a danger in AI running wild. However, the dichotomous thinking that leaves the human teammate solely responsible for all critical and ethical decisions not only leads to unnecessary constraints, but also assumes the human teammate is always superior. Research suggests that 80% of the time that humans fail to take moral action, they actually recognize and understand the more appropriate action to take; these findings suggest a moral judgment-action gap regarding human intentions (Sweeney, Imboden, & Hannah, 2015) . While many humans are aware of their poor moral choices, sometimes they fail to even recognize the moral implications. Humans have limited attentional resources (Kahneman, 1973) and various endurance constraints. Thus, their inability to consistently devote time and awareness to ethical considerations is also a distinct concern; sometimes we simply miss the problem altogether. With so many examples of human teammates actively choosing wrong over right (e.g. Enron, Bill Clinton, David Petraeus), and the inherent limitations of human processing, it is important to challenge the notion of human superiority and consider how, or when, machines may prove optimal. Our overarching argument is that a hybrid human-AI team, when properly configured, will result in superior outcomes. Teams and individuals can achieve these outcomes by either acting independently or as homogeneous teams of each acting without the other.

AI Capabilities Impacting Distributed Team Cognition

Team Data Processing Capabilities

AI can increase the volume, velocity, and consistency of data processing to offer teams an unprecedented cognitive power that improves their information-to-decision capabilities. The main advantage to AI systems is their ability to access and process large volumes of diverse, digital information (Sycara & Lewis, 2004) and to make predictions at unprecedented speed. In recent displays of advanced AI capabilities, such as IBM's Project Debater (Teich, 2018) and Google's Automated Assistant (DeMers, 2018), the latest intelligent systems can sift through vast data repositories to swiftly detect pertinent information, rapidly manage knowledge, and transmit logical arguments and decisions commensurate to their human creators. The ever-increasing volume and velocity of information that is detected, stored, and manipulated through machine learning offers unbounded opportunities to optimize team functionality. When seated with the right communication links, the AI teammate can disperse these cognitive capabilities and optimize team performance from nearly anywhere in the world.

Team Communication Capabilities

Within the realm of distributed team cognition, AI can increase effectiveness by offering a broader, more comprehensive repertoire of cognitive skills. Combining these skills with advanced communication functions can help optimize critical team cognitions, including an increase in shared mental models (Mohammed & Dumville, 2001) and a more robust transactive memory (Wegner, 1987). Paired with human teammates, the rapid and expansive information processing capability of AI systems can help share vital, real-time information to the team while simultaneously serving as a gargantuan, on-demand repository of historical knowledge. Beyond knowledge sharing, AI systems can passively monitor and record team processes or actively push alerts and goal-relevant status updates. This type of information flow can help enrich cue strategies (Salas, Rosen, Burke, & Goodwin, 2009) across a distributed team and ensure a near real-time dissemination of mission information (Salas et al., 2009). These enhanced functions, coupled with the machine-learning features integral to AI systems, offer a powerful situational awareness capability that can be rapidly and consistently shared to all team members. Mitigating the cognitive processing limitations of their human teammates, the AI "cognitive wingman" can also detect and analyze vast amounts of available stimuli to maintain and optimize team shared awareness and simultaneously transmit relevant predictions to all teammates.

Team Situational Awareness Capabilities

AI systems are capable of gathering information at lower levels, while performing information-processing activities to support team coordination and team-level situational awareness (Cooke, Salas, Kiekel, & Bell, 2004). Situational awareness is really about possessing real-time knowledge of what is happening around you (Endsley, 2000; Parasuraman, Sheridan, & Wickens, 2008), or more precisely, "the perception of the elements in the environment within a volume of time and space, the comprehension of their meaning, and the projection of their future status" (Endsley,

1988, p. 97). Endsley (1988) offered three levels of situational awareness: perception, comprehension, and prediction. When furnished with the right sensory capabilities, AI systems can be far more efficient than humans across all three levels to dramatically improve team behavior efficiencies. Specifically, AI can support rapid and thorough situational assessments, continuous situational monitoring, and alerting (Sycara & Lewis, 2004).

We do not expect all AI interpretations to be correct of course, as 100% reliability of autonomous agents cannot be guaranteed (de Visser & Parasuraman, 2011; de Visser et al., 2018). However, we can learn to counter the effects of imperfect performance of AI systems that, compared to human counterparts, have greater behavioral consistencies. Furthermore, we believe imperfections in machine learning are not the real concern. The actual issue is not whether autonomous agents are fallible, but whether or not human teammates are *less* fallible. The gaps in human-machine comparisons are perhaps most salient when evaluating capabilities to detect and interpret stimuli, retrieve and manipulate stored information, or project outcomes to aid in team decision making.

Additionally, researchers have noted that when human and machines work together, substantial performance gains often follow (de Visser & Parasuraman, 2011). Those that dogmatically argue in favor of the human may miss important opportunities to leverage our cognitive "wingman" to increase the consistency, speed, and volume of team situational awareness and improve information-to-decision capabilities. Since exploitation of AI systems can deliver important benefits to distributed team cognition, perhaps it is time to extend the repeated ideology of *keeping the man in the loop* by adding the notion of *keeping machines in the loop*.

In sum, we believe in leveraging AI capabilities to expand the volume of knowledge available to each teammate; improve the velocity of information transmitted throughout the team; and ensure consistency of team processes to vastly improve distributed team cognition. These advanced cognitive capabilities can improve a wide range of team processes, behaviors, and attitudes (see Table 9.1). We argue trustworthy AI systems are uniquely suited to deliver these capability requirements and improve the information-to-decision processes within teams. Specifically, we recommend developing and exploiting AI to deliver the following applied solutions: *enhance critical thinking and adaptive processes*; *improve ethical decision making within teams*; and ultimately, allow teams to *safely exercise disciplined initiative*.

Applied Solution 1: Exploit AI as a Cognitive Support System to Enhance Critical Thinking and Adaptive Processes in Teams

Benefits to Team Critical Thinking

Leveraging AI capabilities as a cognitive support system not only enhances situational awareness but can also help guide critical thinking in teams. It is well known that human thinking is often biased and distorted, yet our productivity and efficiency depend on the very quality of our thought processes (Elder & Paul, 2008). After building a list of common errors[1] that human teammates often make when constructing and evaluating information and arguments, Gerras (2008) argued that we need

TABLE 9.1
AI Capabilities and Potential to Improve Distributed Team Cognition

AI Enhanced Capabilities	**Improved Processes, Behaviors, and Attitudes**
Volume of Information:	***Temporal Processes (Marks, Mathieu, & Zaccaro, 2001):***
• Knowledge repository	• Transition phase
• Knowledge management	• Action phase
• Knowledge transmission	• Interpersonal phase
Velocity of Information processing:	***Common Behaviors (Salas et al., 2009):***
• Detection (sensory)	• Communication (closed-loop, feedback)
• Manipulating	• Coordination
• Storing and classifying	• Cooperation (monitoring, backup)
• Retrieving	***Team Attitudes (Marks et al., 2001;Salas et al., 2009):***
Awareness Levels (Endsley, 1988):	• Efficacy/potency
• Perception	• Emotional regulation
• Comprehension	• Trust
• Projection/prediction	• Goal commitment
Human Error Mitigation:	***Team Cognition (Salas et al., 2009):***
• Avoid endurance constraints	• Cue strategies
• Cognitive load	• Problem solving
• Fatigue	• Mission information
• Mood	• Shared mental models
• Detect and mitigate bias	• Transactive memory
• Cultural insensitivity	• Team situational awareness
• Cognitive/affective bias	***Team Adaptation:***
• Moral intensity perceptions	• Adaptation phases (Burke, Coovert et al., 2006)
• Resist social influence	• Situational assessment
• Groupthink	• Plan formulation
• Group polarization	• Plan execution
• Conformity	• Team learning
• Unethical tendencies	• The *Four Rs* (Frick, Fletcher, Ramsay, & Bedwell, 2018)
• Machiavellianism	• Recognize
• Need for dominance	• Reframe
• Personality disorders	• Respond
	• Reflect

to think more critically about critical thinking itself. Human thought and decision making can suffer from limited cognitive load capacities and fatigue. These human endurance constraints can also be exacerbated through varying shifts in mood, leading to improper judgments or even a general unwillingness to engage in critical thought altogether. Even when the human processing systems are fresh and attentive, cognitive and affective bias can distort perceptions, judgments, and interpretations of environmental stimulus. Furthermore, teams are highly sensitive to social influence and are susceptible to the effects of groupthink (Janis, 1972), group polarization (Myers & Lamm, 1976), and pressures to conform (Milgram, 1974).

AI, in the form our "cognitive wingman." AI can help mitigate these human endurance constraints because they are not susceptible to fatigue or dramatic mood swings. Cognitive activities, typically managed by human teammates, can be offloaded to AI support systems (Cooke et al., 2004) to relieve cognitive pressures across the team. These cognitive support systems are always ready and available to provide a full set of cognitive behaviors to back up, monitor, and reinforce their human teammates. When programmed correctly, they may prove less vulnerable to common biases that contaminate human thought, such as availability and representative heuristics; sample size and regression to the mean bias; overconfidence and arrogance; and finally, confirmation bias and fundamental attribution errors (Gerras, 2008). Although it is possible that human bias can contaminate the algorithms and data inputs necessary to run AI, these cognitive support systems are not directly influenced by social pressures. AI developers could also mitigate bias contamination by building systems that avoid the pitfalls of human cognition. With the right design, AI can guide teams to "think critically about critical thinking" (Gerras, 2008) by performing sensitivity analyses to account for the potential influence of perception bias, variations in cultural norms, and other sources for errors in thought.

Benefits to Adaptive Team Processes

Including AI in teams to improve situational awareness and enhance critical thinking can also help optimize team adaptation skills. Typically, critical thinking and situational awareness alone are not enough to deliver efficient team performance, as team adaptation skills are the crucial characteristics of effective teams (Frick et al., 2018). However, these two cognitive processes are important for improving a team's ability to navigate the team adaptation phases of situational assessment, plan formulation, plan execution, and team learning (Burke, Stagl, Salas, Pierce, & Kendall, 2006). Armed with the impressive volume of knowledge and velocity of information delivered by AI support systems, teams can now use advances in data processing and prediction to avoid the haphazard planning and execution often employed by maladaptive teams. When viewed through the lens of Frick et al.'s (2018) Four R heuristic of *Recognize*, *Reframe*, *Respond*, *and Reflect*, AI systems can provide the situational awareness and critical thinking support to help teams better engage in the following cognitive tasks:

- *Recognize*: Gain information and transmit team-relevant knowledge to ensure all teammates fully understand the current situation and share the same predictions of the future operating environment. Evaluate internal and external cues to identify the core problem that is driving the need for team adaptation.
- *Reframe*: Visualize the desired conditions and identify the capability shortfalls the team can resolve through adaptation. Propose multiple adaptation options, assess available resources, evaluate potential effectiveness, perform sensitivity analyses, and check for bias. Set and disseminate goals and new team roles to build shared mental models.
- *Respond*: Shape the environment and adjust the team to meet the new demands and execute team tasks as necessary to achieve the desired end state. Monitor performance, provide back-up behaviors, and communicate

effectively. Cooperate and coordinate with external teams to synchronize actions.

- *Reflect*: Evaluate the effectiveness of the team adaptation and solidify or refine the new team processes as appropriate.

Applied Solution 2: Apply AI as an Ethical Cognitive Support[2] Tool (Mathieson, 2007)

An Ethical Cognitive Support Tool

Given human temptations to indulge in amoral pursuits (Ludwig & Longenecker, 1993) and our vulnerability to subconscious cognitive bias, perhaps we should use caution when depending on the human as the sole ethical decision maker. Specifically, we should seek to determine which teammate (human or machine) is *more* dangerous and how we might leverage AI to reduce the risk or error involved with human judgment. Applying AI as an ethical cognitive support tool (Mathieson, 2007) has the potential to vastly improve the ethical conduct and effectiveness of teams. Although many might question the ethical implications of AI utilization, there is less discussion about how AI could actually improve organizational ethics within the realms of leadership, climates and cultures, and decision making.

Ethical leadership is believed to produce a trickle-down effect in organizations, as an ethical (or unethical) culture can originate with leaders and cascade directly onto followers within a team (Schaubroeck et al., 2012). These leaders can also indirectly influence the ethical culture of teams across various hierarchical levels in an organization (Schaubroeck et al., 2012). Brown, Treviño, and Harrison (2005) describe ethical leadership as normatively appropriate behavior which can be promoted to followers through two-way communication, reinforcement, and decision making. AI can serve as an important partner in promoting this normatively appropriate behavior for teammates to emulate, especially through gained efficiencies in the communication of ethical codes and the selection of ethical decisions.

Leaders and members of distributed teams often occupy boundary-spanning positions (Drescher, Korsgaard, Welpe, Picot, & Wigand, 2014; Druskat & Wheeler, 2003), and therefore are more likely to encounter ethical dilemmas and ambiguity (Brown et al., 2005) as they coordinate with external teams and encounter unique environmental stimulus. AI support systems could help guide teams through ethical ambiguity by generating various viewpoints, recognizing and transmitting value differences, interpreting ethical codes from different cultures, and predicting unintended consequences of ethical decisions. In fact, most research shows a code of ethics, along with positive ethical cultures and climates, improves ethical decision making (O'Fallon & Butterfield, 2005). Thus, AI support systems can help transmit an organization's code of ethics across distributed teams, while promulgating and reinforcing the normatively appropriate behavior within each teammate's unique contextual environment. This constant awareness of culture and norms can lead to greater ethical team outcomes, as our "cognitive wingman" can ensure teams have the body of knowledge necessary to make informed ethical decisions. If developed

TABLE 9.2
Factors Influencing Perceptions of Moral Intensity

Six Moral Intensity Factors	Description
(1) Magnitude of consequences	Agreement on importance
(2) Social consensus	Event probability × likely effect
(3) Probability of effect	Time between the decision and its consequences
(4) Temporal immediacy	Nearness among decision makers and those affected
(5) Proximity	Intensity of impact by group size
(6) Concentration of effect	Amount of benefit (harm)

Source: Jones, 1991

properly, AI systems could even provide greater insight into implicit perceptions regarding the moral intensity of sensitive issues.

Calculating Moral Intensity Perceptions

When designing and implementing an AI ethical cognitive support system, it is important to recognize and accommodate the six factors (see Table 9.2) that can influence human perceptions regarding moral intensity (Jones, 1991) and whether a decision actually has ethical importance (Mathieson, 2007). Research shows strong support for the idea that perceptions of moral intensity can influence ethical decision making (O'Fallon & Butterfield, 2005). Thus, building AI systems to predict moral intensity perceptions could help guide teams through ethical decision-making processes. By producing and sharing the confidence intervals of these predictions, AI can arm teams with the moral intensity risk associated with various ethical decision-making options. Specifically, our "cognitive wingman" could calculate the probability of effects of various decisions and predict consequences regarding the amount of harm; the degree of social consensus on importance; the amount of time and proximity for an anticipated effect; and even the group size impacted. Given this new depth of information, teams can better navigate basic components of moral decision making (Rest, 1986), which includes the moral nature of issues, moral judgments, moral intent, and moral actions. If AI can help teams take a more deliberate approach in making informed ethical decisions and create a shared awareness, perhaps it is also time build these systems to exercise some basic initiative on behalf of the team.

Applied Solution 3: Deliberately Leveraging AI to Safely Exercise Disciplined Initiative

The Importance of Disciplined Initiative

Deliberately leveraging AI to safely exercise disciplined initiative and accept prudent risk is a critical consideration, especially to the functionality and effectiveness of distributed and crisis action teams. Within a military context, decentralized execution, or what is now often referred to as mission command doctrine

(Department of the Army, 2019), is an important concept for meeting the demands of modern warfare. The mission command doctrine is the US Army's answer to directing and controlling teams, while appropriately empowering decision making and decentralized execution at lower levels. Perhaps now more than ever, war is extremely complex, rapidly changing, and always uncertain (McMaster, 2015). Therefore, a dogmatic approach of centralized control will likely remain too inflexible to capitalize on unforeseen opportunities or to allow teams to quickly adapt during moments of ambiguity. Ever since the German army first began exploring this concept in the late 1800s, the ideas of decentralized leadership and empowering subordinates to take initiative (Matzenbacher, 2018) have repeatedly been proven beneficial to team performance on the battlefield. Now more than ever, modern warfare and the rapid pace of conflict does not allow subordinates to wait for updated orders. Predicated on mutual trust, commanders on the battlefield provide intent and then grant subordinates the flexibility to take critical and decisive action to meet that intent.

Interestingly, the benefits and need for decentralized control is not unique to military organizations. Distributed teams and remote collaboration are becoming increasingly common in our modern work environment. Many teams now work together across vast distances, but often at the same time over a shared visual workspace (Gutwin & Greenberg, 2004). These remote team designs must share a common workplace awareness about the environment and how each physical workspace might change over time. Similar to a military context, these professional work groups will require near real-time awareness of various teammates and how they are interacting with and continuously adapting to their distinct physical workspace. Unfortunately, it is extremely difficult for distributed teams to maintain this necessary awareness (Gutwin & Greenberg, 2004) and to effectively execute team coordination. These limitations lead to process loss and decrements to productivity (Steiner, 1972), along with shortfalls in implicit and explicit team communication (Fiore & Salas, 2004). These gaps can produce detrimental misunderstandings that wreak havoc on distributed team cognition, leaving remote and crisis action teams with a limited ability to adapt and execute decentralized operations or tasks.

Enabling Decentralized Execution

Keeping our cognitive teammate *in the loop* can vastly improve the shared awareness needed for both military teams and professional work groups. Since AI can rapidly transmit information across the team and avoid human endurance constraints, it can continuously monitor and disseminate real-time information relevant to the physical and digital environment. This can increase awareness levels and mitigate the communication and coordination gaps that prevent effective and decentralized execution among distributed teams. Organizations properly enabled with AI can have faster access to mission orders (or goal-based tasks) and an enhanced ability to accept measurable and prudent risk. With greater team situational awareness and rapid information-to-decision capability, teams can develop the cohesion and collective efficacy needed to build mutual trust. AI can help create a shared understanding, disseminate clear intent and evolving goals throughout the team, and calculate how to safely practice decentralized execution (Department of the Army, 2019) of important tasks or missions.

CONCLUSIONS

The goal of our chapter has been to contemplate issues related to the adoption of truly helpful and powerful technologies. These technologies will likely be empowered by AI. The history of technological development is littered with failed products that were produced and, had they been adopted, would certainly have helped society. There are many reasons technologies are not embraced—poor human factors design is likely at the front of that list. Yet, as sociotechnical systems theory so clearly demonstrates, we need to consider human and social issues in addition to technological ones. As such, our chapter has focused not on algorithms and hardware, but rather on issues associated with the human side of the equation. We examined trust and specific issues related to trust in advanced technologies. Following trust, we interwove humanness and the role it plays with the anthropomorphism of software and embodied technologies. Embodied humanness predicts a linear to nonlinear function describing the Uncanny Valley, where increasing degrees of preference for humanness are suddenly replaced by dislike, when the technology becomes *too* human. The third section of our chapter poses our hybrid perspective on when technology will become supremely useful and most adopted. Colloquially stated, our premise is one where we understand the strength and limitations of human cognitions, as well as the strengths and limitations of advanced AI technologies; so, let us combine the two in such a fashion the technology acts as a "cognitive wingman." By providing this type of support, AI can compensate for the bounds on and biases of human cognitive processing.

In developing a premise or theory as to why technology is adopted we need to specify constructs linked in a nomological network encompassing features of the technology, human, and task environment in which the technology is to be deployed. With constructs, such as trust, identified we must then focus on measurement issues to operationalize the premise/theory and begin to examine and test linkages among and between constructs. With a measurement system in place, such as that provided by latent change scores, we have a solid approach to measure trust and how it changes in response to specific aspects of the technology. For example, how much does transparency lead to increased levels of trust? We can explicitly examine the changing levels of trust in the human as we change levels and types of transparency in the technology. Concomitantly, when trust is violated as in the case of false alarms or when the system simply errs, latent change scores will reflect the decrement in trust as a function of those alarms or errors. Finally, we can use the same methodology to evaluate the effectiveness of alternative trust repair strategies.

The issue of appearance on a dimension of humanness for cognitive agents is an important one. We have, essentially, a balancing act where we want greater similarity to humans than that which is presented by the glowing red light of HAL in *2001: A Space Odyssey*; yet many find human-appearing robots such as Erica (Collins, September 13, 2018) creepy and well into the Uncanny Valley. Finding this sweet spot is an important issue for the acceptance of technologies. It is perhaps even more complicated a problem than it may appear on the surface as much research in psychology has demonstrated appearance is correlated with attributions of other characteristics as well. For example, Arbogast (2018) recently demonstrated a significant relationship between appearance of facial features (in both men and women) and an attribution of toxic leadership. In addition to these main effects, individual differences also likely play a role in preferences. As such, it may be most effective to allow individuals to select or even configure the appearance of their personal cognitive wingman in order to maximize acceptance.

As stated above, we believe the cognitive wingman offers the best of both the human world and the technological/AI world. Humans are limited by bounds on their perceptual and cognitive processing systems. We are further limited by faulty heuristics and biases. If we construct AI technologies in such a fashion to augment cognition and reduce cognitive and perceptual limitations while simultaneously overcome biases, the combined human-AI technology team could be quite potent and effective. Even in the relatively benign realm of food processing, providing robots the ability to see has led to a 100% increase in the amount of food processed and other areas have seen a million-fold improvement (Kahn, 2018). To keep these advances coming we must further develop our theories of why technologies are adopted. We have made solid strides in this area by keeping trust as a core construct and examining factors that influence it. For example, minimizing false alarm rates, transparency (of various types), ability to configure critical aspects of the system according to individual difference preferences, and effective strategies for trust repair are key aspects that must be kept in mind. We also need to attend to additional ways to signal trust. This can be done via interface development and the oh so important training programs that can be used to make explicit those factors discussed in this chapter.

What is some of the low-hanging fruit we might consider in the human-cognitive wingman approach? The world continues to change rapidly and we are prone to change blindness and loss of situational awareness. A cognitive wingman could be most helpful in these areas by bringing the problem back into focus for the human. As described previously, critical thinking is one of the most important aspects of cognition done well. A cognitive wingman-human team can provide results superior than can be achieved by either alone. Of course, whenever teamwork is involved it is critical to monitor and maximize team functioning to ensure process gain over process loss. A well-constructed cognitive wingman, in addition to providing task-specific knowledge and expertise, can also monitor the process between members of the team, including the AI members and, when appropriate, act as facilitators ensuring effective team process.

In closing, let is point out the essential importance of ethical AI and how it may ensure ethical behavior among humans as well. The essence of ethical behavior in our society is not always clearly defined; there are many gray areas. Yet in order to construct an ethical AI system we need to really grapple with what it means to be an ethical human. Doing this hard work of defining the criterion space of the construct, providing examples for training so the system can have codified knowledge, and developing a learning system which itself understands ethical behavior, are minimally three components that would serve as the kernel to such an ethical AI system. Grappling with the development of these three components would force us to profoundly consider what it means to be ethical in our society today. This process will help us manage the current ethical problems faced by humans and will ensure ethical behavior from AI in the role it plays as our cognitive wingman.

NOTES

1. The nine most common errors reflecting weak critical thinking are (Gerras, 2008): arguments against the person, false dichotomies, appeals to unqualified authorities, false causes, appeals to fear, appeals to the masses, slippery slope arguments, weak analogies, and red herrings.
2. Early discussions of AI envisioned its applicability for decision support. Our perspective is the utility of AI is broader, encompassing many diverse cognitive activities. As such we employ the term "cognitive support."

REFERENCES

Alarcon, G. M., et al. (2017). A descriptive model of computer code trustworthiness. *Journal of Cognitive Engineering and Decision Making, 11*, 107–121.

Arbogast, M. S. (2018). *Egos gone wild: Threat detection and the domains indicative of toxic leadership.* Unpublished dissertation. Tampa, FL: University of South Florida Department of Psychology.

Army, U. S. (2019). *Army Doctrine Publication (ADP) No. 6–0, mission command: Command and control of army forces.* Washington, DC: US Department of Defense.

Asimov, I. (1942). Runaround. *Astounding Science Fiction, 29*(1), 94–103.

Bostrom, N., & Yudkowsky, E. (2014). The ethics of artificial intelligence. *The Cambridge Handbook of Artificial Intelligence, 1*, 316–334.

Broadbent, E., Kumar, V., Li, X., Sollers, J., Stafford, R. Q., MacDonald, B. A., & Wegner, D. M. (2013). Robots with display screens: a robot with a more humanlike face display is perceived to have more mind and a better personality. *PLoS One, 8*(8), e72589. https://doi.org/10.1371/journal.pone.0072589

Brown, M. E., Treviño, L. K., & Harrison, D. A. (2005). Ethical leadership: A social learning perspective for construct development and testing. *Organizational Behavior and Human Decision Processes, 97*(2), 117–134.

Burke, C. S., Stagl, K. C., Salas, E., Pierce, L., & Kendall, D. (2006). Understanding team adaptation: A conceptual analysis and model. *Journal of Applied Psychology, 91*(6), 1189–1207.

Burke, J., Coovert, M. D., Murphy, R., Riley, J., & Rodgers, E. (2006). Human-robot factors: Robots in the workplace. *Proceedings of the annual meeting of the Human Factors and Ergonomics Society* (pp. 870–874). San Francisco, CA: Sage.

Byrne, E. A., & Parasuraman, R. (1996). Psychophysiology and adaptive automation. *Biological psychology, 42*(3), 249–268.

Clifford, C. (2018). *Steve Wozniak explains why he used to agree with Elon Musk, Stephen Hawking on A.I.—but now he doesn't.* Retrieved from www.cnbc.com/2018/02/23/steve-wozniak-doesnt-agree-with-elon-musk-stephen-hawking-on-a-i.html

Collins, P. (2018, September 13). Creepy Japanese robot "with a soul" will replace a human news anchor. *The Daily Good.* Retrieved from www.good.is/articles/robot-soul-anchor.

Cooke, N. J., Salas, E., Kiekel, P. A., & Bell, B. (2004). Advances in measuring team cognition. In E. Salas & S. Fiore (Eds.), *Team cognition: Understanding the factors that drive process and performance* (pp. 83–107). Washington, DC: APA Books.

Coovert, M. D., Lee, T., Shindev, I., & Sun, Yu. (2014). Spatial augmented reality as a method for a mobile robot to communicate intended movement. *Computers in Human Behavior, 34*, 241–248.

Coovert, M. D., Miller, E. E. P., & Bennett, W. (2017). Assessing trust and effectiveness in virtual teams: Latent growth curve and latent change score models. *Social Sciences, 6*(3), 87.

Delgado, M., Frank, R., & Phelps, E. (2005). Perceptions of moral character modulate the neural systems of reward during the trust game. *Nature Neuroscience, 8*, 1611–1618.

DeMers, J. (2018). Google assistant is getting better: Here's what that means for marketers. *Forbes*. Retrieved from www.forbes.com/sites/jaysondemers/2018/05/29/google-assistant-is-getting-better-heres-what-that-means-for-marketers/#20547fe8616b

de Visser, E. J., Krueger, F., McKnight, P., Scheid, S., Smith, M., Chalk, S., & Parasuraman, R. (2012). The world is not enough: Trust in cognitive agents. *Proceedings of the Human Factors and Ergonomics Society Annual Meeting, 56*, 263–267. https://doi.org/10.1177/1071181312561062

de Visser, E. J., Pak, R., & Shaw, T. H. (2018). From "automation" to "autonomy": The importance of trust repair in human–machine interactions. *Ergonomcs, 61*(10), 1409–1427. https://doi.org/10.1080/00140139.2018.1457725.

de Visser, E. J., & Parasuraman, R. (2011). Adaptive aiding of human-robot teaming: Effects of imperfect automation on performance, trust, and workload. *Journal of Cognitive Engineering and Decision Making, 5*(2), 209–231.

Drescher, M. A., Korsgaard, M. A., Welpe, I. M., Picot, A., & Wigand, R. T. (2014). The dynamics of shared leadership: Building trust and enhancing performance. *Journal of Applied Psychology, 99*(5), 771.

Druskat, V. U., & Wheeler, J. V. (2003). Managing from the boundary: The effective leadership of self-managing work teams. *Academy of Management Journal, 46*(4), 435–457.

Dzindolet, M., Peterson, S., & Pomranky, R. (2003). The role of trust in automation reliance. *International Journal of Human-Computer Studies, 58*(6), 697–718.

Elder, L., & Paul, R. (2008). Critical thinking: The nuts and bolts of education. *Optometric Education, 33*(3).

Endsley, M. R. (1988). Design and evaluation for situation awareness enhancement. In *Proceedings of the Human Factors Society annual meeting* (Vol. 32, No. 2, pp. 97–101). Los Angeles, CA: SAGE Publications.

Endsley, M. R. (2000). Theoretical underpinnings of situation awareness. In M. R. Endsley & D. J. Garland (Eds.), *Situation awareness and measurement* (pp. 3–32). Boca Raton, FL: CRC Press.

Epley, N., Waytz, A., & Cacioppo, J. (2007). On seeing human: A three-factor theory of anthropomorphism. *Psychological Review, 114*, 864–886.

Feigh, K. M., Dorneich, M. C., & Hayes, C. C. (2012). Toward a characterization of adaptive systems: A framework for researchers and system designers. *Human Factors, 54*(6), 1008–1024.

Fiore, S. M., & Salas, E. E. (2004). Why we need team cognition. In E. E. Salas & S. M. Fiore (Eds.), *Team cognition: Understanding the factors that drive process and performance* (pp. 235–248). Washington, DC: American Psychological Association.

Fraser, N. M., Hipel, K. W., Kilgour, D. M., McNeese, M. D., & Snyder, D. E. (1989). An architecture for integrating expert systems. *Decision Support Systems, 5*(3), 263–276.

Frick, S. E., Fletcher, K. A., Ramsay, P. S., & Bedwell, W. L. (2018). Understanding team maladaptation through the lens of the four R's of adaptation. *Human Resource Management Review, 28*(4), 411–422.

Galford, R. M., & Drapeau, A. S. (2003). The enemies of trust. *Harvard Business Review, 81*(2), 85–95.

Gees-Blair, K., Rice, S., & Schwark, J. (2013). Using system-wide trust theory to reveal the contagion effects of automation false alarms and misses on compliance and reliance in a simulated aviation task. *International Journal of Aviation Psychology*, *23*, 245–266.

Gerras, S. J. (2008). Thinking critically about critical thinking: A fundamental guide for strategic leaders. *Carlisle, Pennsylvania: US Army War College*, *9*.

Good, I. J. (1965). Speculations concerning the first ultraintelligent machine. *Advances in Computers*, *6*(99), 31–83.

Gray, K., & Wegner, D. M. (2012). Feeling robots and human zombies: Mind perception and the uncanny valley. *Cognition*, *125*(1), 125–130.

Gray, K., Young, L., & Waytz, A. (2012). Mind perception is the essence of morality. *Psychological Inquiry*, *23*(2), 101–124. https://doi.org/10.1080/1047840X.2012.651387

Gutwin, C., & Greenberg, S. (2004). The importance of awareness for team cognition in distributed collaboration. In E. E. Salas & S. M. Fiore (Eds.), *Team cognition: Understanding the factors that drive process and performance* (pp. 177–201). Washington, DC: American Psychological Association.

Ho, C., & MacDorman, K. F. (2010). Revisiting the uncanny valley theory: Developing and validating an alternative to the Godspeed indices. *Computers in Human Behavior*. Retrieved from http://linkinghub.elsevier.com/retrieve/pii/S0747563210001536

Ho, C.-C., MacDorman, K. F., & Pramono, Z. A. D. D. (2008). Human emotion and the uncanny valley. In *Proceedings of the 3rd international conference on Human robot interaction—HRI '08* (p. 169). New York: ACM Press. https://doi.org/10.1145/1349822.1349845

Janis, I. L. (1972). *Victims of groupthink: A psychological study of foreign-policy decisions and fiascoes*. Boston, MA: Houghton Mifflin.

Jones, T. M. (1991). Ethical decision making by individuals in organizations: An issue-contingent model. *Academy of Management Review*, *16*(2), 366–395.

Kaber, D. B. (2018a). A conceptual framework of autonomous and automated agents. *Theoretical Issues in Ergonomics Science*, *19*(4), 406–430.

Kaber, D. B. (2018b). Issues in human—automation interaction modeling: Presumptive aspects of frameworks of types and levels of automation. *Journal of Cognitive Engineering and Decision Making*, *12*(1), 7–24.

Kaber, D. B., & Endsley, M. (2004). The effects of level of automation and adaptive automation on human performance, situation awareness and workload in a dynamic control task. *Theoretical Issues in Ergonomics Science*, *5*(2), 113–153.

Kahn, N. (2018, September 12). Seefood: Why robots are producing more of what you eat. *Wall Street Journal*. Retrieved from www.wsj.com/994b4773-d72c-47e3-9091-d440428ba204.

Kahn, P. H., Severson, R. L., Kanda, T., Ishiguro, H., Gill, B. T., Ruckert, J. H., . . . Freier, N. G. (2012). Do people hold a humanoid robot morally accountable for the harm it causes? In *Proceedings of the seventh annual ACM/IEEE international conference on human-robot interaction—HRI '12* (p. 33). New York, NY: Association for Computing Machinery. https://doi.org/10.1145/2157689.2157696

Kahneman, D. (1973). *Attention and effort* (Vol. 1063). Englewood Cliffs, NJ: Prentice-Hall.

Kim, P. H., Dirks, K. T., & Cooper, C. D. (2009). The repair of trust: A dynamic bilateral perspective and multilevel conceptualization. *Academy of Management Review*, *34*(3), 401–422.

Kim, T., & Hinds, P. (2006, September). Who should I blame? Effects of autonomy and transparency on attributions in human-robot interaction. In *ROMAN 2006-The 15th IEEE international symposium on robot and human interactive communication* (pp. 80–85). Piscataway, NJ: IEEE.

Krueger, F., McCabe, K., Moll, J., Kriegeskorte, N., Zahn, R., Strenziok, M., . . . Grafman, J. (2007). Neural correlates of trust. *Proceedings of the National Academy of Sciences*, *104*(50), 20084–20089.

Kurzweil, R. (2005). *The singularity is near: When humans transcend biology*. New York: The Viking Press.

Lee, J., & See, K. (2004). Trust in automation : Designing for appropriate reliance. *Human Factors*, *46*, 50–80.

Lewicki, R. J., & Brinsfield, C. (2017). Trust repair. *Annual Review of Organizational Psychology and Organizational Behavior*, *4*, 287–313. https://doi.org/10.1146/annurev-orgpsych-032516-113147

Ludwig, D. C., & Longenecker, C. O. (1993). The Bathsheba syndrome: The ethical failure of successful leaders. *Journal of Business Ethics*, *12*(4), 265–273.

Lyons, J. B. (2013). Being transparent about transparency: A model for human-robot interaction. In *Trust and autonomous systems: Papers from the 2013 AAAI Spring symposium* (pp. 48–53). Boston, MA: AAAI Press.

Lyons, J. B., Ho, N. T., Koltai, K. S., Masequesmay, G., Skoog, M., Cacanindin, A., & Johnson, W. W. (2016). Trust-based analysis of an Air Force collision avoidance system. *Ergonomics in Design*, 9–12. https://doi.org/10.1177/1064804615611274

Lyons, J. B., Koltai, K. S., Ho, N. T., Johnson, W. B., Smith, D. E., & Shively, R. J. (2016). Engineering trust in complex automated systems. In *Ergonomics in design, January, 13–17.* Boston, MA: Sage.

MacDorman, K. F. (2005). Subjective ratings of robot video clips for human likeness, familiarity, and eeriness: An exploration of the uncanny valley. In *ICCS/CogSci-2006 long symposium: Toward social mechanisms of android science*. Retrieved from www.macdorman.com/kfm/writings/pubs/MacDorman2006SubjectiveRatings.pdf

MacDorman, K. F., & Ishiguro, H. (2006). The uncanny advantage of using androids in cognitive and social science research. *Interaction Studies*, *7*(3), 297–337. https://doi.org/10.1075/is.7.3.03mac

MacDorman, K. F., & Kageki, N. (2012). The uncanny valley: The original essay by Masahiro Mori. In *IEEE spectrum*. New York, NY: IEEE Press.

Madhavan, P., & Wiegmann, D. (2007). Similarities and differences between human-human and human-automation trust: An integrative review. *Theoretical Issues in Ergonomics Science*, *8*(4), 277–301.

Marks, M. A., Mathieu, J. E., & Zaccaro, S. J. (2001). A temporally based framework and taxonomy of team processes. *Academy of Management Review*, *26*, 356–376.

Mathieson, K. (2007). Towards a design science of ethical decision support. *Journal of Business Ethics*, *76*(3), 269–292.

Matzenbacher, M. B. (2018, March–April). The US Army and mission command. *Military Review*, 2018, pp. 61–71.

May, J. F., & Baldwin, C. L. (2009). Driver fatigue: The importance of identifying causal factors of fatigue when considering detection and countermeasure technologies. *Transportation Research Part F: Traffic Psychology and Behaviour*, *12*(3), 218–224. https://doi.org/10.1016/j.trf.2008.11.005.

McAllister, D. J. (1995). Affect-and cognition-based trust as foundations for interpersonal cooperation in organizations. *Academy of Management Journal*, *38*(1), 24–59.

McMaster, L. G. H. (2015, March–April). The army operating concept and clear thinking about future war. *Military Review*, 2015, pp. 6–21.

McNeese, M. D. (1986). Humane intelligence: A human factors perspective for developing intelligent cockpits. *IEEE Aerospace and Electronic Systems*, *1*(9), 6–12.

Milgram, S. (1974). *Obedience to authority: An experimental view*. New York: Harper & Row.

Mohammed, S., & Dumville, B. C. (2001). Team mental models in a team knowledge framework: Expanding theory and measurement across disciplinary boundaries. *Journal of Organizational Behavior: The International Journal of Industrial, Occupational and Organizational Psychology and Behavior*, *22*(2), 89–106.

Mori, M. (1970). The uncanny valley. *Energy, 7*(4), 33–35. Tokyo: Nihon Enerugi Gakkaishi/ Journal of the Japan Institute of Energy.

Myers, D. G., & Lamm, H. (1976). The group polarization phenomenon. *Psychological Bulletin, 83*(4), 602.

Nass, C., Moon, Y., Fogg, B., Reeves, B., & Dryer, D. (1995). Can computer personalities be human personalities? *International Journal of Human-Computer Studies, 43*(2), 223–239.

Nass, C., Steuer, J., & Tauber, E. (1994). Computers are social actors. In *CHI '94 Proceedings of the SIGCHI conference on Human factors in computing systems* (pp. 73–78). Boston, MA: ACM.

O'Fallon, M. J., & Butterfield, K. D. (2005). A review of the empirical ethical decision-making literature: 1996–2003. *Journal of Business Ethics, 59*(4), 375–413.

Parasuraman, R., Sheridan, T., & Wickens, C. (2000). A model for types and levels of human interaction with automation. *IEEE Transactions on Systems, Man and Cybernetics, Part A: Systems and Humans, 30*, 286–297.

Parasuraman, R., Sheridan, T., & Wickens, C. (2008). Situation awareness, mental workload, and trust in automation: Viable, empirically supported cognitive engineering constructs. *Journal of Cognitive Engineering and Decision Making, 2*, 140–160.

Rest, J. R. (1986). Moral development: Advances in research and theory. Retrieved from http://hdl.handle.net/10822/811393

Rousseau, D. M., Sitkin, S. B., Burt, R. S., & Camerer, C. (1998). Not so different after all: A cross-discipline view of trust. *Academy of Management Review, 23*, 393–404.

Salas, E., Rosen, M. A., Burke, C. S., & Goodwin, G. F. (2009). The wisdom of collectives in organizations: An update of the teamwork competencies. In *Team effectiveness in complex organizations: Cross-disciplinary perspectives and approaches* (pp. 39–79). New York, NY: Psychology Press.

Sanfey, A. G., Rilling, J. K., Aronson, J. A., Nystrom, L. E., & Cohen, J. (2003). The neural basis of economic decision-making in the Ultimatum Game. *Science, 300*(5626), 1755–1758.

Saygin, A. P., Chaminade, T., Ishiguro, H., Driver, J., & Frith, C. (2011). The thing that should not be: Predictive coding and the uncanny valley in perceiving human and humanoid robot actions. *Social Cognitive and Affective Neuroscience, 7*(4), 413–422. https://doi.org/10.1093/scan/nsr025.

Scerbo, M. W. (1996). Theoretical perspectives on adaptive automation. *Automation and Human Performance: Theory and Applications*, 37–63.

Scerbo, M. W. (2008). Adaptive automation. *Neuroergonomics: The Brain at Work, 3*, 239.

Schaubroeck, J. M., Hannah, S. T., Avolio, B. J., Kozlowski, S. W., Lord, R. G., Treviño, L. K., . . . Peng, A. C. (2012). Embedding ethical leadership within and across organization levels. *Academy of Management Journal, 55*(5), 1053–1078.

Seyama, J., & Nagayama, R. S. (2007). The uncanny valley: Effect of realism on the impression of artificial human faces. *Presence: Teleoperators and Virtual Environments, 16*(4), 337–351. Retrieved from www.mitpressjournals.org/doi/abs/10.1162/pres.16.4.337

Snyder, D. E., & McNeese, M. D. (1987). *Conflict resolution in cooperative systems* (No. AAMRL-TR-87-066). Wright-Patterson AFB, OH: Harry G Armstrong Aerospace Medical Research Lab.

Steckenfinger, S. A., & Ghazanfar, A. A. (2009). Monkey visual behavior falls into the uncanny valley. *Proceedings of the National Academy of Sciences of the United States of America, 106*(43), 18362–18366. https://doi.org/10.1073/pnas.0910063106

Steiner, I. D. (1972). *Group process and productivity*. New York: Academic Press.

Sweeney, P. J., Imboden, M. W., & Hannah, S. T. (2015). Building moral strength: Bridging the moral judgment-action gap. *New Directions for Student Leadership, 146*, 17–33. https://doi.org/10.1068/p6900

Sycara, K., & Lewis, M. (2004). Integrating intelligent agents into human teams. In E. E. Salas & S. M. Fiore (Eds.), *Team cognition: Understanding the factors that drive process and performance* (pp. 203–231). Washington, DC: American Psychological Association.

Teich, D. A. (2018). IBM research project debater: Closer to passing the turing test. *Forbes*. Retrieved from www.forbes.com/sites/davidteich/2018/06/26/ibm-research-project-debater-closer-to-passing-the-turing-test/

Thompson, J. C., Trafton, J. G., & McKnight, P. (2011). The perception of humanness from the movements of synthetic agents. *Perception, 40*, 695–704. https://doi.org/10.1068/p6900

Wegner, D. M. (1987). Transactive memory: A contemporary analysis of the group mind. In *Theories of group behavior* (pp. 185–208). New York: Springer.

Wellens, A. R. (1993). Group situation awareness and distributed decision making: From military to civilian applications. In J. Castellan (Ed.), *Individual and group decision making: Current Issues* (pp. 267–291). Hillsdale, NJ: Lawrence Erlbaum Associates.

Zak, P. J. (2017). The neuroscience of trust. *Harvard Business Review, 95*(1), 84–90.

Index

A

C

D

E

F

G

H

I

J

K

L

M

N

O

P

R

S

T

W